Agricultural Planning, Technology and Management

Agricultural Planning, Technology and Management

Edited by **Nancy Cahoy**

SYRAWOOD
PUBLISHING HOUSE

New York

Published by Syrawood Publishing House,
750 Third Avenue, 9th Floor,
New York, NY 10017, USA
www.syrawoodpublishinghouse.com

Agricultural Planning, Technology and Management
Edited by Nancy Cahoy

International Standard Book Number: 978-1-68286-007-6 (Hardback)

The publisher's policy is to use permanent paper from mills that operate a sustainable forestry policy. Furthermore, the publisher ensures that the text paper and cover boards used have met acceptable environmental accreditation standards.

Trademark Notice: Registered trademark of products or corporate names are used only for explanation and identification without intent to infringe.

Printed in the United States of America.

Contents

Preface

Agriculture is one of the most extensively practiced economic activity worldwide. It is also responsible for fulfilling the world demand for food security. This has accelerated the research in this sector. Scientists and researchers across the globe are working to device innovative and sustainable agricultural practices. Some of the varied topics covered in this elaborate book are analysis of technical efficiency, cost structures, effect of drought, etc. This book targets students as well as agriculture scientists, researchers, policy makers and professionals engaged in the field of agriculture at various levels.

The researches compiled throughout the book are authentic and of high quality, combining several disciplines and from very diverse regions from around the world. Drawing on the contributions of many researchers from diverse countries, the book's objective is to provide the readers with the latest achievements in the area of research. This book will surely be a source of knowledge to all interested and researching the field.

In the end, I would like to express my deep sense of gratitude to all the authors for meeting the set deadlines in completing and submitting their research chapters. I would also like to thank the publisher for the support offered to us throughout the course of the book. Finally, I extend my sincere thanks to my family for being a constant source of inspiration and encouragement.

Editor

Analysis of the Effects of Agricultural Inputs Price Liberalization on the Production of Sunflower in Khoy Zone

Ali Bagherzadeh[1] and Fatemeh kazemzadeh[2]*

Abstract

Sunflower is one of four main annual oil plants that cultivated in oil and nut varieties. This plant as an important and industrial food product and because of nutritional features and the potential for earning exchange has become a valuable product in foreign and inner markets and has a special position in agricultural sector. Khoy, by producing 40 percent of sunflower productions of country annually, is the greatest sunflower producer in Iran. The main purpose of this study is the analysis of the effects of inputs price liberalization on production of sunflower producers in this city. This study is according to a field research and cross-sectional data of 2009 have been used for it. Results show input price liberalization policy by increasing inputs prices and decreasing demand amounts of inputs, increases the production costs and decreases the production and totally it's harmful for sunflower producers. For preventing negative effects of liberalization on production, adopting necessary policies such as merging small farms and making big ones to profit by economies of scale and increasing production and productivity with the resulted incomes from liberalization and spending them in scientific researches to produce with low costs are suggested.

Keywords:
Sunflower, Liberalization,
Production factor, Production
elasticity, Khoy

[1] *Assistant professor of Economics, Khoy Branch, Islamic Azad University, Iran.*
[2] *MA in Economics, Tabriz Branch, Islamic Azad University, Iran.*
* *Corresponding author's email: Bagherzadeh_eco58@yahoo.com*

INTRODUCTION

Oil seeds as industrial plants are one of main and strategic products of agricultural sector that are cultivated for providing eatable oil. The importance and value of oil seeds are not only for their oil but also for valuable material that is consumed for nutrition after oil-pressing. Sunflower is one of four main annual oil plants that cultivated in oil and nut varieties. Nut varieties have special oil acids and less oil but oil varieties have 43-49 % oil that is full of D and E vitamins and has an important role in human health. Sunflower oil is important in clearing vessels and preventing brain and heart strokes. Also, it is used to make medicine, soap, colors and cosmetic materials. Because of desirable quality of oil and desirable reaction in unsuitable environment conditions, sunflower has a special position in agricultural sector and can be effective in economies of most countries (Mehrabi et al., 2009).

Khoy is the greatest sunflower producer in Iran that produces 40% of sunflower production in country. Moderate weather, susceptible land, suitable market, having high price in comparison with other agricultural products, being precocious and possibility for second cultivation are the most important reasons of sunflower cultivation in this area. Nut varieties of Khoy are rare in color, taste and size and are competitive with other countries products. The products are exported to Persian Gulf countries and if export way becomes paved, thousands of tons of this production can be exported to other countries (Bagherzadeh,2010).

As we know, agriculture is the economic heart of most countries and most likely source of significant economic growth (DFID, 2003). It has been observed as the major and certain path to economic growth and sustainability. In spite of the dominant role of the petroleum sector as the major foreign exchange earner, agriculture remains the mainstay of the economy (NEEDS, 2004) as the economic of most developing countries are built on agriculture. There is strong relationship between agricultural productivity growth and reduction of poverty. Sunflower in agricultural sector of Iran as an important and

industrial food product and because of nutritional features and the potential for earning exchange has become a valuable product in foreign and inner markets. Many governments intervene, directly in producing agricultural products through taxations and subsidization, so inputs price are not real. For increasing the ability of economy sector, it`s necessary to use clear and competitive prices of inputs to have economic efficiency. Liberalization policy of agricultural sector as a way of development has been proposed to developing countries by World Bank. Liberalization simply means allowing market forces of demand and supply to determine what to provide, for whom to produce, and the method of production to be used in an economy. Liberalization involves deregulation and the removal or reduction of government`s participation in the economy. Their justifying reasons for this policy are environment protection, decreasing government costs, increasing inputs productivity and stable development of agricultural sector. Since this policy has been enforced in our country since some years ago and now continues, there are some concerns about loss of this sector. The analysis of input price liberalization effects on production is too important because enforcing price liberalization policy by increasing inputs prices and decreasing their consumption affects production and production costs and if these effects are negative, they can make some problems that threat economic and political security of country.

Previous studies on input consumption in agricultural sector and subsidies effects show that subsidy elimination has been one of government`s important economic policies in recent decades. In most countries there were such experiences that we imply some of them.

Gulati (1990) in a study on agricultural input price liberalization claimed according to high and increasing costs of agricultural input subsidies, eliminating these subsidies is necessary but it has some reactions on agricultural sector that needs to more study and analysis. Ready and Deshpande (1992) by analyzing the effects of fertilizer price liberalization in India showed that this policy as an agricultural development tool has both positive and negative effects. Elyasian and Hosseini (1996) in a research on the analysis of effects of agricultural inputs subsidy elimination showed in case of wheat profitability after liberalization is twice as much as before liberalization. In another research Azizi (2005) studied the price liberalization effects of poison and fertilizer inputs on rice in Guilan. Results showed that fertilizer was using in second area –economic area- and price liberalization led to increase in price, decrease in fertilizer consumption and so decrease in production, but poison was using in third area, price increase led to consume in second area. Then by comparing the positive and negative effects of liberalization, continuing the policy was suggested for this input. In another study Karimzadegan (2006) presented eliminating fertilizer subsidy decreases it`s consumption for wheat and returning to optimal input consumption increases their production and profit. Ayinde *et al.,* (2009) by studying the effects of fertilizer policy on production in Nigeria in two periods (before liberalization and after liberalization) showed despite fertilizer price increase, liberalization leads to production increase. In this study for having more production, controlling inputs prices and educating farmers are suggested. Badmus (2010) by doing the same research on corn production in Nigeria by using of SUR method on time series of 1970-1998 claimed liberalization is ineffective on fertilizer price and consumption but has positive effects on production. Mousavi *et al.,* (2010) studied the welfare effects of eliminating fertilizer subsidy on corn production in Fars. Results showed liberalization despite price increase did not affect fertilizer demand and so it caused high production costs and low profit.

The main objectives of this study are the analysis of input price liberalization effects on sunflower production and it`s production costs. So according to our objectives, we try to derive production function, cost function, production elasticity of inputs and the comparative importance of inputs to show whether the farmers behave logical in applying inputs or not. Then by means of inputs demand functions and inputs demand price elasticity, consumption and production changes are identified.

MATERIALS AND METHODS

In developing countries including Iran, better use of agricultural inputs like land, fertilizer, poison, water and so on, for increasing the production and development of agricultural sector has special importance and there are several tools to achieve them. One of the most important tools for choosing suitable approaches in production and optimal allocation of sources is using production functions. In fact, production functions are one of analysis ways of quantitative relations between the amounts of inputs and production operation. These functions are mathematical relations that identify input conversion rate to output. So we have:

$$y = f(x_1, x_2, \dots, x_n) \qquad (1)$$

Y is the amount of production and x_is are production inputs.

Agricultural production function is:

$$y = f(x) = Ax_1^\alpha x_2^\beta, \quad \alpha, \beta > 0 \qquad (2)$$

Total cost function is:

$$TC = w_1 x_1 + w_2 x_2 \qquad (3)$$

Lagrange function is:

$$\min: w_1 x_1 + w_2 x_2 \quad subject\ to\ Ax_1^\alpha x_2^\beta \geq y \qquad (4)$$

$$l = w_1 x_1 + w_2 x_2 + \lambda\left(y - Ax_1^\alpha x_2^\beta\right) \qquad (5)$$

For minimizing agricultural sector costs, we make derivation toward each production input and put them equal to zero (Badmus, 2010).

$$\frac{\partial l}{\partial x_1} = w_1 - \lambda\alpha\, f(x)/x_1 = 0 \implies w_1 x_1 = \lambda\alpha f(x) \qquad (6)$$

$$\frac{\partial l}{\partial x_2} = w_2 - \lambda\beta\, f(x)/x_2 = 0 \implies w_2 x_2 = \lambda\beta f(x) \qquad (7)$$

$$w_1 x_1 + w_2 x_2 = \lambda(\alpha+\beta)f(x) \qquad (8)$$

Then we give joint powers to the both sides of first condition equations, find the amount of λ and put in cost equation as follows:

$$w_1^\alpha x_1^\alpha = \lambda^\alpha \alpha^\alpha f(x)^\alpha, \quad w_2^\beta x_2^\beta = \lambda^\beta \beta^\beta f(x)^\beta \qquad (9)$$

$$\implies \left(w_1^\alpha w_2^\beta\right)f(x) = A\lambda^{\alpha+\beta}\left(\alpha^\alpha \beta^\beta\right)f(x)^{\alpha+\beta} \qquad (10)$$

$$= \left(\frac{w_1^{\alpha} w_2^{\beta} f(x)}{A \alpha^{\alpha} \beta^{\beta} f(x)^{\alpha+\beta}} \right)^{\frac{1}{\alpha+\beta}} \quad (11)$$

$$c(y, w_1, w_2) = w_1 x_1 + w_2 x_2 = \lambda(\alpha + \beta) f(x) \quad (12)$$

$$\lambda(\alpha + \beta) f(x) = (\alpha + \beta) B A^{\frac{-1}{\alpha+\beta}} y^{\frac{1}{\alpha+\beta}} \left(w_1^{\frac{\alpha}{\alpha+\beta}} w_2^{\frac{\beta}{\alpha+\beta}} \right)$$

$$, B = \left(\alpha^{\frac{-\alpha}{\alpha+\beta}} \beta^{\frac{-\beta}{\alpha+\beta}} \right) \quad (13)$$

$lnC = ln(\alpha+\beta)+lnB - (1/(\alpha+\beta)) lnA + (1/(\alpha+\beta)) lny + (\alpha/(\alpha+\beta)) ln\ w_1 + (\beta/(\alpha+\beta)) ln\ w_2$ (14)

Coefficients of each parameter in above equation are cost elasticities of inputs. Also we have this result as scale elasticity:

$$\frac{AC}{MC} = \frac{Costs}{y(\partial C/\partial y)} = \left(\frac{\partial \ln C}{\partial \ln y} \right)^{-1} = (\alpha + \beta) \quad (15)$$

Then we find demand functions of each input:

$$x_1(y, w_1, w_2) = \frac{\partial C}{\partial w_1} = \alpha B A^{\frac{-1}{\alpha+\beta}} y^{\frac{1}{\alpha+\beta}} w_1^{\frac{\alpha}{\alpha+\beta}-1} w_2^{\frac{\beta}{\alpha+\beta}} \quad (16)$$

$$x_2(y, w_1, w_2) = \frac{\partial C}{\partial w_2} = \beta B A^{\frac{-1}{\alpha+\beta}} y^{\frac{1}{\alpha+\beta}} w_1^{\frac{\alpha}{\alpha+\beta}} w_2^{\frac{\beta}{\alpha+\beta}-1} \quad (17)$$

This study is according to a field research and cross-section data of 2009 have been used for it. We spent 2 months for collecting data. At first we took the names of all sunflower producers of Khoy that were 5000 units then we classified the statistical universe of producers to 3 classes according to the size of cultivated land, (0-4] hectares, (4-8] hectares and (8, more) hectares, that 60% of lands were between (0.25-4) hectares, 30% of lands were between (4-8) hectares and less than 9% were more than 8 hectares. After classification, by using of classified random sampling, we could get optimal sample volume by:

$$n = \frac{z_{\frac{\alpha}{2}}^2 . \sigma^2}{d^2} \quad (18)$$

So for every class by random method we chose a volume of $n_h = n.N_n/N$ that n is optimal amount of sample for statistical universe. Since our method was random, for identifying optimal n, we should know the sum of all classes is equal to the optimal amount of sample of statistical universe. We know for collecting samples of n>30, mean sampling distribution is normal. So:

$$\sum_{i=1}^h n_h = n \qquad d = Z_{\frac{\alpha}{2}} . \frac{\sigma_x}{\sqrt{n}} \quad (19)$$

$$\sqrt{n} = \frac{Z_{\frac{\alpha}{2}}.\sigma}{d} => n = \left(\frac{Z_{\frac{\alpha}{2}}.\sigma}{d} \right)^2 = \frac{z_{\frac{\alpha}{2}}^2.\sigma^2}{d^2} \quad (20)$$

In above equations, d is the maximum amount of authorized error, σ_x^2, the variance of statistical universe and z, standard normal distribution by α significance level. For calculating variance, we used, $\sigma = (Xmax-Xmin)/\varepsilon$. The variance of statistic universe was 400, so $\sigma_x=20$. Then we calculated the optimal amount of each class.

$$n = \frac{(1.96)^2.400}{(4.65)^2} \approx 70 \quad (21)$$

After having total optimal number and optimal number of sample for each class, by using of random numbers table, we took the names of sample sunflower producers and finally found essential data for study.

RESULTS AND DISCUSSIONS

Empirical studies and sampling methods in producing sunflower seeds show factors like seed, labor, fertilizer and watering times are the main factors of production function in this sector. Applied production function in this study is Cobb-Douglas. Cobb-Douglas function has better goodness of fit toward other agricultural production functions. So this function was used for estimating the relation between sunflower production and inputs. Sunflower production function that has been estimated by ordinary least squares method is presented as:

$lny=-4.092+0.785\ lns + 0.365\ lnl + 0.751\ lnw + 0.139\ lnf$ (22)

t- statistics: (-4.25) (2.9) (1.58) (2.67) (1.85)

The numbers inside parentheses are t- ratios that show all variables of production function

Table 1: The results of estimation by Microfit 4.1

Regressor	coefficient	t-statistic
Constant	-4.092	-4.25
Lns	0.785	2.9
Lnl	0.365	1.58
Lnw	0.751	2.67
Lnf	0.139	1.85
R-squared		0.75
R-Bar-squared		0.69
F-statistic		12.256
DW-statistic		1.8

have significance in 95% confidence level. $R^2=75\%$ and $\bar{R}^2=69\%$ that according to cross-section data are acceptable amounts. Also F-statistic related to analysis of variance is 12.256 that presents the significance of total regression in 95% confidence level. Also according to the results of White test, there is no problem of heteroscedasticity and not having correlation are admitted by (D.W) statistics. Following table presents the results of estimation by Microfit 4.1.

As we know estimated coefficients of equation are inputs production elasticities. According to coefficients, sensitivity of production toward watering times and seed in comparison with others are more and production has the least sensitivity toward fertilizer. All production elasticities are between zero and one that show inputs are used in second area-economic area-that is true about Cobb-Douglas functions. Production elasticity shows the sensitivity of production toward inputs demand. Here seed and water have more sensitivity toward their demand change and fertilizer has the least sensitivity.

In Cobb-Douglas production function, the sum of all production elasticities is the amount of scale elasticity. We have:

$$E=E_s+E_l+E_w+E_f \qquad (23)$$

So scale elasticity is 2.04 that is more than one and shows increasing returns to scale. Increasing returns to scale happens in decreasing

part of long term average cost curve with economies of scale. On the other hand, in this amount, firm can decrease it`s average cost by increasing it`s production.

For studying price elasticities and cost elasticities, we need cost function. According to duality rule, we derive cost function as follows:

$$\ln C = 1.91 + 0.49 \ln y + 0.385 \ln w_1 + 0.179 \ln w_2 + 0.368 \ln w_3 + 0.068 \ln w_4 \qquad (24)$$

$w_i s$ are the price of each inputs. Now we can get demand equations.

$$x_1\,(y,\,w_1,\,w_2,\,w_3,\,w_4) = 1.31.\,0.017^{-0.49}\,y^{0.49}\,w_1^{-0.615}\,w_2^{0.179}\,w_3^{0.368}\,w_4^{0.068} \qquad (25)$$

$$x_2\,(y,\,w_1,\,w_2,\,w_3,\,w_4) = 0.61.\,0.017_{-0.49}\,y_{0.49}\,w_1^{0.385}\,w_2^{-0.821}\,w_3^{0.368}\,w_4^{0.068} \qquad (26)$$

$$x_3\,(y,w_1,w_2,w_3,w_4) = 1.25.0.017^{-0.49}\,y^{0.49}\,w_1^{0.385}\,w_2^{0.179}\,w_3^{-0.632}\,w_4^{0.068} \qquad (27)$$

$$x_4\,(y,w_1,w_2,w_3,w_4) = 2.230.017^{-0.49}\,y^{0.49}\,w_1^{0.385}\,w_2^{0.179}\,w_3^{0.368}\,w_4^{-0.932} \qquad (28)$$

If we write logarithmic form of demand equations, gained coefficients of inputs prices show inputs price elasticities. Following table presents price elasticity amounts of sunflower inputs.

Above table shows price elasticities are completely according to demand rule. All inputs price elasticities are negative. Seed has the least and fertilizer has the most amount of elasticity. 1% increase in seed price lead to 0.615% decrease in seed demand also if fertilizer price increases 1%, it`s demand will decrease 0.932%. Elasticity of fertilizer demand is about one and price changes have more effect on it`s demand. Seed demand because of low severity toward price changes takes less effect. Also this matter is true about other inputs.

Now, if price liberalization policy is enforced,

Table 2: price elasticity and intersecting elasticity of sunflower inputs

Inputs	Seed	Labor	Water	Fertilizer
Seed	-0.615	0.179	0.368	0.068
Labor	0.385	-0.821	0.368	0.068
Water	0.385	0.179	-0.632	0.068
Fertilizer	0.385	0.179	0.368	-0.932

the price of seed and fertilizer will increase. According to demand price elasticity of inputs, farmers demand for these inputs will decrease. On the other hand, with these amounts of production elasticities for inputs, by decreasing their demand, the amount of production will decrease. Of course by using of input substitution, we can take the production level at the same amount. So input price liberalization policy by increasing the price of inputs, increases the costs of production and decreases the production and totally it`s harmful for sunflower producers.

CONCLUSION AND RECOMMENDATION

Study results show that input price liberalization policy has negative effect on sunflower production. According to gained results, all inputs are used in second area of production –economic area- so liberalization policy by increasing inputs prices, according to demand rule, decrease the demand amounts of inputs and decrease in inputs consumption cause production decrease.

Gained price elasticity is about one for fertilizer that shows 1% increase in fertilizer price causes 1% decrease in input demand amount. Decrease in it`s consumption lead to production decrease that is harmful for sunflower producers. According to demand inelasticity of seed, sunflower producers reaction toward price increase won`t be too severe but it will have demand decrease of this input and finally according to seed production elasticity, the amount of production will decrease. In conclusion, the effect of inputs price liberalization policy on sunflower industry, is increase in production costs and decrease in sunflower production. So for enforcing policy, we need adopting exact and planned policies. According to gained results, following suggestions are presented:

Agricultural organizations should have true supervision on the amount, time and the way of input consumption and help farmers to increase their scientific informing to use inputs optimally and therefore increase their production and efficiency.

Government can compensate extra costs of farmers resulting from liberalization policy by giving cash subsidies. These subsidies could be paid based on the amount of production or cultivated land. In this way farmers distribute the cash subsidy among all inputs and decrease severe use of one input.

For preventing negative effects of liberalization on production by merging small farms and making big ones, we can benefit of economies of scale and increase our profit. It`s clear that increase in profit leads to increase in production and efficiency.

Also government should pay more attention to scientific and especially genetic researches and increase the research budget of R&D centers and universities to produce with low costs and improve total factor productivity.

REFERENCES

1- Ayinde, O.E., Adewumi, M.O., & Omotosho, F.J. (2009). Effect of Fertilizer Policy on Crop Production in Nigeria, The Social Sciences. (Abstract).

2- Azizi, J. (2005). The Analysis of Effects of Poison and Fertilizer Price Liberalization on Rice Production in Guilan, Agricultural Economy & Development. (Abstract).

3- Bagherzadeh, A. (2010), Comparative Analysis of Economic-Social Features on Technical Efficiency of Sunflower Farms (The case of Khoy sunflower), Agricultural Sciences of Islamic Azad University of Khoy.

4- Badmus, M.A. (2010). Market Liberalization and Maize Production in Nigeria. Journal of Development and Agricultural Economics. (Abstract).

5- DFID (2003). A DFID policy paper. Htpp://www.dfid.gov.uk

6- Elyasian, H., & Hosseini, M. (1996). Liberalization Effects on Agricultural Inputs Usage, Agricultural Economy & Development. (Abstract).

7- Gulati, A, (1990). Fertilizer Subsidy, Is the Cultivator Net Subsidized? Indian Journal of Agricultural Economics. (Abstract).

8- Karimzadegan, H. (2006). The Effect of Fertilizer Subsidy on Unoptimal Consumption of It in Wheat Production, Agricultural Economy & Development, 14, 55

9- Mehrabi, H., & Pakravan, M. (2009). Calculating the Varieties of Efficiency and Returns to Scale of Sunflower Producers in Khoy, Economy & Agricultural Development. (Abstract) (In Persian).

10- Mousavi, A., & Farajzadeh, Z. (2010). Welfare Effects of Fertilizer Subsidy Elimination on Corn Consumers and Producers, Agricultural Economy

Researches, 2(2): 34-50.

11- National Economic Empowerment and Development Strategy (NEEDS). The NEEDS Secretariat, National Planning Commission, Federal Secretariat

12- Ready, V.R., & Deshpande, R, S. (1992). Input Subsidies: Whither the Direction of Policy Changes, Indian Journal of Agricultural Economics. (Abstract).

2

The Impact of Bio-Ethanol Conversion and Global Climate Change on Corn Economic Performance of Indonesia

Yudi Ferrianta[1], Nuhfil Hanani[2], Budi Setiawan[2] and Wahib Muhaimin[2]

Abstract

Keywords:
The energy crisis, Climate change, Corn

Many studies conclude that the rise in global food prices due to higher demand from the development of bio-fuels, climate anomalies, and increased of oil prices. Not only the food commodity index rose more than 60 percent, non-food commodity price index also rose over 60 percent and crude oil price index has increased even further above 60 percent. The purpose of this study is to analyze the impact of bio-ethanol conversion and global climate change on corn economic performance of Indonesia. The results showed that the food crisis caused by climate anomalies lead the world corn prices rose 50 percent, impact on Indonesia corn imports fell by 11.86 percent. And the other hand, the energy crisis that caused the corn used as feedstock for ethanol that caused U.S. corn exports only 20 percent of their products have an impact on Indonesia on maize imports fell 32.4 percent.

[1] *Department of Agriculture Economic, University of Lambung Mangkurat, Indonesia.*
[2] *Department of Agricultural Economic, University of Brawijaya, Indonesia.*
* *Corresponding author's email: ferrianta@gmail.com*

INTRODUCTION

Corn is the third largest crop after wheat and rice, most of the corn products are used and traded as feed material in addition to a staple food. In addition to food and feed, corn has a wide range of industrial applications such as materials for the manufacture of ethanol.

Over the last decade of global corn production has shown increasing growth, the global corn market generally divided into two issues, first, the conversion of the global corn used as bio-ethanol industry, second, the share of globally traded corn are relatively constant.

The main cereal market - corn, wheat and rice - has shown some major adjustments in recent years. Since 2008 the global food crisis resulted in a large spike in corn prices. On the demand side of high oil prices encourage the development of bio-fuel which resulted in increased demand in addition to dietary changes and income and population growth. High oil prices also put pressure upward on the cost of crop production (e.g. fertilizer, tillage). On the supply side with low cereal stocks, exacerbated by a policy of trade restrictions on cereal and speculation in commodity markets. (Flammini, 2008).

The food crisis followed by the global financial crisis in the second half of 2008, high oil prices which led to concerns about the security of national oil and concerns about the environmental impact of fossil fuel use resulting in searching alternative energy sources, one of the interesting issues is the development of bio-fuels that affect the global corn market.

In the United States, the enhanced production of bio-ethanol because corn prices are relatively low, In the year 2007-2008, as many as 82 million tones of corn used for ethanol, which represents a quarter of U.S. corn production and 12% of global production (DEFRA, 2008). Besides the development of bio-ethanol, one of the factors that cause serious problems for the production of corn from time to time is the occurrence of El Niño weather phenomenon associated with an abnormal warming of sea surface temperatures in the Pacific Ocean. Corn plants are most affected by El Niño (mostly in the form of prolonged dry conditions) are con-centrated in the southern hemisphere, particularly in southern Africa. During El Niño events of the 1980s and the 1990s, for example, corn production in the Republic of South Africa fell by 40 to 60 percent. Also in Brazil, corn producers suffered from floods and droughts driven by El Niño situation in the past. adverse weather conditions caused by the events of the last major El Niño of 1997/98 are located mostly in East Asia and led to a sharp decline in production in countries such as Thailand and Indonesia.

In Indonesia, corn has a very strategic role, especially for the farm development and other industries. In past, corn mainly used as staple. However, currently, corn mainly used as an industrial material. In line with the rapid growth of livestock industry, it is estimated more than 55% of domestic corn needs is used for feed, while for food consumption is only about 30%, and the remainder for other industrial needs and seeds (Indonesia Department of Agriculture, 2010).

Currently, the development of corn production can not meet high demand. Therefore, the governments meet the shortage of these needs through imports. For 2010 forecast figures, with area of 3 million ha of crops, it is estimated to produce 12.1 million tons.. Meanwhile, maize demand in the country reached 13.8 million tons, resulting in a shortage about 1 million ton to be imported (Ferrianta, 2012). If the import increment increase was not controlled, it will cause a reduction in foreign exchange, and can lower the domestic maize price, where the price was relatively low. Based on these facts, the government is trying to meet the domestic maize need through maize self-sufficiency program.

Maize self-sufficiency effort must be directed to external factors, not only change in domestic policy but also external shock e.g bio-ethanol development and global climate change. In line with the development of world economy, maize commodity will face a different environment. External and internal shock will affect corn economic performance of Indonesia.

Based on these facts, it is deemed necessary to conduct research on the impact of bio-ethanol development and global climate change on the economic performance of corn in Indonesia.

MATERIALS AND METHODS

Indonesia maize economic model is a simultaneous equations consisting of three sub-models: sub production, sub domestic market and sub world markets. The data collected is secondary time series data.

Model estimation is done using re-specification model. The goal is to obtain good models based on economic and econometrics criteria. In the estimation of these models studied the problem of identification, aggregation and the degree of correlation between explanatory variables.

Evaluation conducted to know the impact of instrument change simulation variable on the future endogenous variable. The evaluation model is based on economic theory and information related to the research phenomenon. A model is good if it meets the following criteria:

1. Economics, in association with signs and estimation parameters,

2. Statistics, relating to statistical tests, and

3. Econometrics, related to the model assumptions (Baltagi, 2008)

For unbiased and consistent estimations, simultaneous systems require a more complex procedure for estimation than single equation models, which can generally be estimated by regression with ordinary least squares (OLS). The most frequently used method of estimating simultaneous systems is the two-stage least squares (2SLS) method (Studenmund, 1997; Greene, 1993).

Furthermore, because the model contains a simultaneous equations and lagged endogenous variables, serial correlation test is performed using statistical dw (Durbin-Waston Statistics) in each equation. (Gujarati, 2004)

Model validation performed to analyze how constructed model could to represent the real world. In this study, statistical validation criteria for value estimate econometric model is Root Means Squares Error (RMSE), Root Means Squares Percent Error (RMSPE) and Theil's Inequality Coefficient.

Econometric modeling and estimation can be useful in providing a retrospective look at the economic effects of a policy change or external shock (MCDaniel, 2006). To simulation the impact of external shock to the import corn Indonesia, this study was used ex-post econometrics analysis to see changes in the value of endogenous variable due to changes in exogenous variables. (T. B. Palaskas 1988 ; Baumann, 2011).

Dynamic simultaneous equations system used to develop econometric model. Models specification used are described as follows:

1. $QJ = AJ * PRJ$

2. $AJ = a_1 PJ + a_2 Pkdl_{t-1} + a_3 AJ_{t-1} + U_1$

3. $PRJ = b_1 Pp + b_2 i + b_3 AJ + b_4 W + b_5 PRJ_{t-1} + b_6 CH + U_2$

4. $DIT = DIP + DIL + DK$

5. $DIP = c_1 Ppk + c_2 Pj + c_3 Pkdl + c_4 DIP_{t-1} + U_3$

6. $DIM = d_0 + d_1 Pop + d_2 PJ + d_3 Pni + U_4$

7. $DK = e_0 + e_1 PJ + e_2 Y + e_3 DK_{t-1} + U_5$

8. $MIT = MIAS + MICH + MITH + MIO$

9. $MIAS = f_1(PIAS - PIAS_{t-1}) + f_2 QJ + f_3 DIT + f_4 ERI + f_5(RISTI - RISTI_{t-1}) + U_6$

10. $MICH = g_1 PICH + g_2 QJ + g_3 DIT + g_4 RISTI + U_7$

11. $MITH = h_1 PITH + h_2 QJ + h_3 DIT + h_4 RISTI + U_8$

12. $MIO = MIT - (MIAS + MICH + MITH)$

13. $RISTI = (PJ - PWJ)/ PWJ$

14. $PIAS = PWJ + RISTAS$

15. $PICH = PWJ + RISTCH$

16. $PITH = PWJ + RISTTH$

17. $PJ = i_1 MIT + i_2 DIT + U_9$

18. $XAS = j_0 + j_1 QAS + j_2 DAS + j_3 XTH + j_4 XCH + j_5 MJJ + j_6 MJK + j_7 PETH + U_{10}$

19. $XCH = k_1 QCH + k_2 DCH + j_3 XAS + j_4 XTH + j_5 MJJ + j_6 MJK + U_{11}$

20. $XTH = l_0 + l_1 PWJ + l_2 QTH + l_3 DTH + U_{12}$

21. $MJJ = m_0 + m_1 PWJ + m_2 NPRJj + m_3 ERj + U_{13}$

22. $MJK = n_0 + n_1 PWJ + n_2 DJk + n_3 MJK_{t-1} + U_{14}$

23. $XW = XAS + XTH + XCH + XRO$

24. $MW = MJJ + MJK + MRO$

25. $PW = o_1 XW + o_2 MW + U_{15}$

Note:

• AJ = acreage of corn harvested (ha)

• PRJ = productivity corn of Indonesia (tones / ha)

• QJ = corn production of Indonesia (tones)

• PJ = corn prices of Indonesia(US $ / tone)

• i = Indonesia interest rate (%)

• W = Indonesia wage labor (US $ / day)

• Pp = the price of fertilizer (US $/ tone)

• CH = climate change (oceanic nino index)

- DIT = total corn demand of Indonesia (tones)
- DIP = Indonesia corn demand for feed industry (tones)
- DIM = Indonesia corn demand for food industry (tones)
- DK = Indonesia corn demand for direct consumption (tones)
- KDP = feed prices of Indonesia (US $ / tone)
- Pkdl = soybean price of Indonesia (US $ / tone)
- Pop = population of Indonesia (people)
- MIT = Total Imports corn of Indonesia (tones)
- MIAS = Indonesia corn imports from US. (tones)
- MICH = Indonesia corn Import from China (tones)
- MITH = Indonesia corn imports from Thailand (tones)
- MIO = Indonesia corn imports from other countries (the rest)
- PIAS = the price of corn imports from US (US $ / ton)
- PITCH = the price of corn imports from China (US $ / ton)
- PITH = the price of corn imports from Thailand, (US $ / ton)
- RISTI = corn trade restrictions of Indonesia
- ERI = exchange rate of Indonesia (rupiah / US $)
- XAS = US corn exports (thousand tones)
- XTH= Thailand corn exports (thousand tones)
- XCH= Chinese corn exports (thousand tones)
- QAS = U.S. corn production (thousand tones)
- QTH = Thailand corn production (thousand tones)
- QCH= Chinese corn production (thousand tones)
- MJJ = Japan corn imports (thousand tones)
- PET = ethanol price (US$/bushel)
- MJK = Korea corn imports (thousand tones)
- DJ = corn demand of Korea (thousand tones)
- NPRJ= corn trade restrictions of Japanese (thousand tones)
- ER = exchange rate of Japan (Yuan / US $)
- XW = world exports (thousand tones)
- XRO = corn exports of other country (thousand tones)
- MJW = world corn imports (thousand tones)
- MRO =corn exports of other country (thousand tones)

This study used *time series data's*, starting in 1983 until 2010. The data is obtained from Indonesia Department of Agriculture, Bureau of Indonesia Statistics, Indonesia Ministry of Agri-culture, Directorate General of Food Crops and Horticulture, Food and Agriculture Organization, United States Department of Agriculture, United Nations Commodity Trade Statistics Database, and International Monetary Fund.

RESULTS

International corn economy has undergone major changes over the past two decades in terms of production, utilization, trade and marketing structure. This change was driven by a number of factors ranging from rapid advances in seed technology and production, changes in national policy and international trade, expansion almost without interruption from the use of feed throughout the world and recently huge demand for ethanol.

Production

Over the past two decades, global corn production has increased nearly 50 percent, or 1.8 percent growth rate per year. Most of the increase in world corn production over the past decade can be attributed to rapid expansion in Asia.

Asian corn production grew nearly 35 percent over the past decade, nearly 30 percent of the global increase. The increasing expansion of acreage and yield contributed to high growth rates, like China that makes the most significant progress with contributions as much as 60 percent of total corn production of Asia over the past decade.

Although progress is associated with varieties that have high productivity, it is likely to increase corn production in many countries remains large along with the good level of production efficiency, especially in developing countries are still under

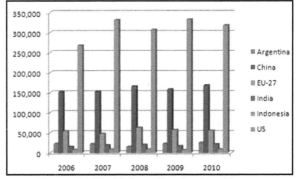

Figure 1: World Corn Production (Sources: USDA, 2010)

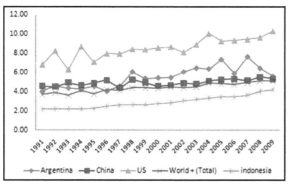

Figure 2: Corn Productivity in Some Countries
(Sources: FAOSTAT, 2010)

major manufacturers. Average corn yields among developing countries about a third of the countries major corn producer. Among some of the countries largest producer of corn (Figure 2), Argentina approximately 5.6 tones / ha, China about 5 tones / ha, while Indonesia about 3 tones / ha. This is much compared to the United States about 10 tones / ha.

Corn as Biofuels Material

Bio-diesel is an alternative diesel fuel energy sources are derived from vegetable oil (vegetable oil) and animal fats (animal fat) where corn is one potential source of bio-diesel product. World price of bio-diesel (FOB Central Europe) increased to $ 4.14 per gallon by 2010, driven by high oil prices and prices of edible oils. Increased crude oil prices and the existence of tariff barriers in Argentina, Brazil, European Union, as well as the U.S. led to an increase in world prices.

With the huge consumption demand for local industrial production of bio-diesel will increase production by 5% in 2010 and estimated production continues to increase and reached 3.5 billion gallons by 2019, on the other hand the consumption continues to grow to 4.0 billion gallons by 2019 so that the net import grow during the outlook period and reached 559 million gallons by 2019.

General Estimation of Econometrics Model

The empirical result of prediction models in the study is good. All exogenous variables included in the structural model has a parameter that the sign suitable with the theory and logical. Statistical criteria used in evaluating the prediction

is quite good. Coefficient of determination (R^2) value in each behavioral equations ranged from 0,38 to 0.99. From 15 behavioral equations, there is only one behavioral equation with R^2 values of 31 percent and 14 other equation is above 64 percent. This shows that, in general, the exogenous variables included in the structural equation model can explain variance rightly for each endogenous variable.

The value of statistic F test generally high. There are 12 of 15 equation had value greater than 11.22. Meanwhile, only two equations have F-value 8,50 and a 1,38. That is, simultaneously, explanatory variable variance in each equation behavior are able to explain the variance of endogenous variable, at $\alpha = 0.0001$; $\alpha = 0.0003$ and $\alpha = 0.2744$. Detailed econometric model estimation for maize are presented in Table 1.

Simulation of External Shocks on the Economic Performance of Indonesia corn Performance

The world has been experiencing a global crisis caused by global warming, energy crises, and monetary crisis. Global warming has caused climate anomaly, resulting in a sharp decline in world agricultural production resulting food crises, including maize.

Global food price index increase has reached 120 percent, where about 60 percent in just the past two years, while the World Bank stated that the price index of food crops increased 86 percent between 2006 to 2008. Agricultural commodity prices rose in 2006 and 2007 and continued to increase even more sharply in 2008. Meanwhile, according to the World Bank, global wheat prices increased by 81 percent (World Bank, 2008), and 83 percent increase in overall global food prices.

The energy crisis has led to the development of corn as a bio-fuel feedstock, resulting in a decrease in world corn exports, especially in the US. The figure below shows the extent of the use of corn for the bio-fuels industry the United States, 1984-2009 a huge surge in the use of corn as an ethanol feedstock domestic product, this indication will be a large drop in exports US, in addition to other major exporting

Table 1: Econometrics Model Estimation

Model	Variable	coefficient	t-statistic	statistic
AJ	PJ	5.3656	0.41	0.6882
	PKDLL	-0.37924	-0.75	0.4618
	AJL	0.962391	7.61	<.0001
	F-test= 1016.30	R²= 0.99284	DW = 1.73635	
PRJ	PUPUK	-5.87E-10	-2.57	0.0186
	I	-0.04547	-3.01	0.0072
	AJ	5.259E-08	0.31	0.7584
	W	-0.00018	-0.41	0.6869
	PRJL	1.246336	14.97	<.0001
	CH	-0.02339	-1.21	0.2416
	F-test = 2640.65	R²= 0.9988	DW = 1.286932	
DIP	PPK	41.06936	2.19	0.0397
	PJ	-34.4608	-1.15	0.2642
	DIPL	0.881228	4.48	0.0002
	PKDL	-0.33793	-0.17	0.8647
	F-test= 92.47	R²= 0.94628	DW = 1.241174	
DIM	Intercept	-15280000	-3.86	0.0009
	PJ	-128.926	-1.13	0.2709
	PNI	56.18753	5.59	<.0001
	POP	0.023537	3.18	0.0045
	F-test=12.62	R²= 0.64325	DW = 0.880593	
DK	Intercept	769281.5	1.18	0.2496
	PJ	-22.447	-0.94	0.3578
	Y	-62.2421	-0.45	0.654
	DKL	0.652367	7.14	<.0001
	F-test= 58.46	R²= 0.89306	DW = 0.985704	
DIT	DIT = DIP + DIM + DK			
MIAS	PIASH	-50301.5	-0.32	0.7514
	QJ	-0.39767	-5.13	<.0001
	DIT	0.403125	5.33	<.0001
	ERI	-16.0265	-0.84	0.4118
	RISTIH	-75.6419	-0.21	0.8352
	F-test= 13.82	R²= 0.77557	DW = 2.307239	
MICH	PICH	-540.767	-0.35	0.728
	QJ	-0.15933	-1.27	0.2183
	DIT	0.195036	1.9	0.0716
	RISTI	-624179	-1.83	0.0815
	F-test= 8.28	R²= 0.61206	DW = 1.874842	
MITH	PITH	-819.626	-1.62	0.1206
	QJ	-0.02801	-0.73	0.4737
	DIT	0.046871	1.41	0.1737
	RISTI	-45811	-0.37	0.7169
	F-test= 8.76	R²= 0.62518	DW = 1.429704	
MIT	MIAS + MICH+ MITH+MIO			
PJ	MIT	-0.00093	-0.3	0.7638
	DIT	0.001993	6.89	<.0001
	F-test= 86.26	R²= 0.88237	DW = 0.136375	
XAS	Intercept	9262248	0.11	0.9148
	QAS	0.061165	0.53	0.5999
	DAS	-0.02246	-0.14	0.8919
	XTH	-0.03033	-0.01	0.991
	XCH	-0.53669	-1.57	0.1352
	MJJ	3.226394	0.61	0.5487
	MJK	0.971053	0.91	0.3738
	PETH	-23310000	-1.38	0.1848
	F-test= 1.52	R²= 0.38465	DW = 1.937323	
XCH	QCH	0.174571	1.2	0.2436
	DCH	-0.23146	-1.26	0.2232
	XAS	-0.23542	-1.76	0.0948
	XTH	-0.74319	-0.66	0.517
	MJJ	1.194996	2.27	0.0348
	MJK	0.638035	1.02	0.3217
	F-test=11.23	R²=0.78010	DW = 2.12621	
XTH	Intercept	-427222	-1.98	0.0614
	PWJ	2148.176	2.73	0.0124
	QTH	1.011109	20.58	<.0001
	DTH	-0.93046	-40.62	<.0001
	F-test= 612.58	R²= 0.98870	DW = 2.079181	
XW	XAS +XTH +XCH + XO			
MJJ	Intercept	21603517	10.37	<.0001
	PWJ	-8992.84	-1.44	0.165
	NPRJ	-148722	-1.49	0.1504
	ERJ	-24401.7	-4.73	0.0001
	F-test=14.61	R²=0.67603	DW = 2.337765	
MJK	Intercept	-122641	-7.82	<.0001
	PWJ	-234.022	-1.85	0.0789
	DJK	1.005725	536.15	<.0001
	MJKL	0.003463	2.03	0.0549
	F-test= 130684	R²= 0.99995	DW = 1.885428	
MW	MJK + MJJ + MIT +MJO			
PW	XW	-1.41E-07	-0.18	0.8566
	MW	1.72E-06	2.2	0.0382
	F-test= 227.42	R²= 0.95187	DW =1.126065	

Source: Research findings.

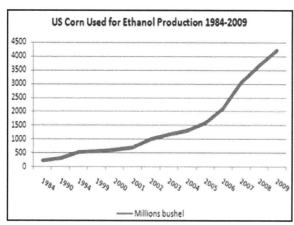

Figure 3: US Corn used for Ethanol Production 1984-2009 (Sources: USDA, 2010)

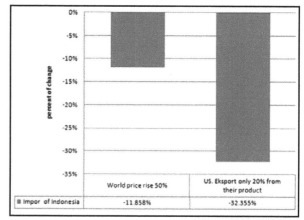

Figure 4: Estimated of Indonesia Corn Import at The Event of External Shock (Simulation analysis)

countries, especially in the European Union and Latin America jointly develop bio-ethanol industry. (Figure 3).

Simulation aims to analyze the impact of various changes in the exogenous variables. However, before doing the simulation, model validation must be done to look at the suitability of the predicted value in accordance with the actual value of each endogenous variable (Pindyck and Rubinfield, 1991).

Table 2 presents the results of the validation of the economic corn model. Based on Table 2 can be found, only three equations in the model has a RMSPE value of more than 50 percent, only one equation is greater than 100 percent and the rest have RMSPE value of less than 50 percent. U-Theil criteria there are 13 equations

have a U value of less than 0.20, and 5 the equation has a value of U between 0.24 to 0.50. The highest value of the Theil-U in the equation is 0.5, and RMPSE value greater than 100 percent, is owned by the Indonesian corn price equation but there is no systematic bias, because the value of Um more than 0.20. Overall, this model is suitable for use as predictive models, so the structural model has been formulated which can be used for various simulations.

Simulation is used with the assumptions: (1) climate anomalies lead the world corn prices rose 50 and the energy crisis that caused the corn used as feedstock for ethanol, as a result the world corn prices rose 2.9 percent. Ex ante analysis for simulation model presented in table 3. Based

Table 2: Result of Validation Dynamic Econometric Models

No.	Variable	RMSPE	Reg (UR)	Var (US)	Covar (UC)	Coef U
1	AJ	0.5904	0.02	0.04	0.76	0.0028
2	PRJ	11.6864	0.02	0.02	0.00	0.0558
3	DIP	1.8582	0.01	0.00	0.37	0.0089
4	DIL	22.8999	0.00	0.00	0.00	0.1293
5	DK	38.9767	0.05	0.06	0.00	0.2509
6	MIAS	52.9878	0.99	0.62	0.37	0.2636
7	MICH	65.7784	0.25	0.11	0.14	0.4613
8	MITH	24.6001	0.14	0.00	0.14	0.1445
9	PJ	262.4	0.07	0.00	0.07	0.5018
10	PWJ	37.3681	0.08	0.08	0.00	0.2411
11	XAS	7.4399	0.22	0.22	0.00	0.0399
12	XTH	30.5445	0.10	0.10	0.00	0.1909
13	XCH	21.8008	0.00	0.00	0.00	0.1226
14	MJJ	3.0589	0.07	0.01	0.06	0.0150
15	MJK	0.3193	0.08	0.08	0.00	0.0016
16	DIT	16.9178	0.01	0.01	0.00	0.0926
17	QJ	12.0027	0.03	0.03	0.01	0.0574
18	MIT	30.1795	0.59	0.18	0.41	0.1761

Source: Research findings.

Table 3: The Ex-Post Analysis for Simultaneous Simulation

Variable	Base	World price rise 50%	US. Export only 20% from their product
AJ	2156142	2156584	2157446
PRJ	5.4348	5.4348	5.4348
DIP	5338632	5335799	5330258
DIL	6838667	6828066	6807336
DK	842862	841017	837408
MIAS	345879	338761	0
MICH	150932	103884	141042
MITH	119420	52425.8	117113
PJ	24919.9	25002.2	25162.9
PWJ	160.9	240	165.6
XAS	52286034	50281511	9242391
XTH	534673	704465	544645
XCH	6605682	6090199	16681027
MJJ	17055710	16344914	17013965
MJK	9501793	9483296	9500706
DIT	13020161	13004881	12975002
QJ	11747862	11750272	11754655
MIT	1106718	985558	748642

Source: Research findings.

on analysis can be show that the food crisis due to climate anomalies lead the world corn prices rose 50 percent impact on Indonesia corn imports fell by 11.86 percent. While the energy crisis that caused the corn used as feedstock for ethanol, causes U.S. limit maize exports only 20 percent of their products have an impact on corn imports Indonesia fell 32.4 percent. (Figure 4)

Ex-post simulation analysis results indicate that both the external shock of climate change and bio-ethanol conversion have an impact the decline in the amount of corn traded in world markets, and this has also impacted on the decline in imports of corn Indonesia. Based on the decrease in the value from the two simulations shows that the U.S. has a significant role in Indonesian corn economy, where if the U.S. lowered its export causes a considerable impact for more decline in Indonesia imports. This indicates Indonesia has a large dependence on the U.S for maize domestic supply, therefore the need for policy in terms of increased productivity and the expansion of planting area by utilizing the technology package and the existing land use to reduce dependence on other countries.

CONCLUSION AND RECOMMENDATIONS

The main objective of this study was to knows the impact of bio-ethanol development and global climate change on the economic performance of corn in Indonesia. This study used the annual time series data (1983-2010) and use a dynamic simultaneous equations system.

Ex- post simulation analysis results show that climate change, bio-ethanol conversion as a major determinant of import corn Indonesia. The results of simulation analysis of global climate change and the conversion of corn for bio-ethanol have an impact on the fall to import corn in Indonesia. This situation is expected to increase competitive advantage and comparative of Indonesian corn farming. Nevertheless there are still many problems faced by Indonesia such as corn farm land issues, technology, human resources, capital, fertilizer; rural infrastructure; and distortion distribution.

Some policies that are needed include (1) the expansion of planting area by increasing cropping index (IP) and extensification by making use of idle land, (2) suppress the difference in results between regions and agro-ecosystems through the use of new high yielding varieties and hybrid composites as well as site-specific application of the PTT model, (3) suppress the loss of the harvest and post harvest, and (4) increase the stability of the results between seasons and regions through the implementation of integrated pest management wisely. (5) human resource development of farmers through Farmer Field Schools (human capital) and also involve farmers in innovation (joint innovation), (6) institutional development (social capital) farmers as Farmer Field School activities continued; (7) irrigation infrastructure investment and drainage are more flexible (physical capital), and (8) investment in infrastructure and rural economy.

REFERENCES

1- Baltagi, B.H. (2008). Econometrics. Springer Publishing, Germany.

2- Mcdaniel,Ch., reinert, K., & Hughes, K. (2006). Tools of the Trade:Models for Trade Policy analysis. Science, Technology, America, and the Global Economy Woodrow Wilson International Center for Scholars. Pennsylvania Avenue Washington. DC.

3-Department for Environment Food and Rural Affairs (DEFRA). (2008). The Impact of Biofuels on Commodity Prices. Nobel House. London

4- Ferrianta, Y. (2012). Impact of Trade Liberalization Asean-China Free Trade Area (ACFTA) on The Performance of Indonesia Maize Economy. (Dissertation). University of Brawijaya Malang. Indonesia.

5-Flammini, A. (2008). Biofuels and the Underlying Causes of High Food Prices. Rome, Global Bioenergy Partnership.

6-Greene, W.H. (1993). Econometric Analysis (2nd ed.), New York: Macmillan.

7-Gujarati, D.N. (2004). Basic Econometrics, 4th ed. McGraw-Hill, New York.

8- Indonesia Department of Agriculture (2010). Prospects and Corn Development in Indonesia. Agency for Agricultural Research and Development, Ministry of Agriculture, Jakarta.

9- Pindyck, R.S & Rubinfeld, D.L. (2005). Economic Model and Economic forecasts. McGraw Hill, Inc. New York.

10- Baumann, R., & Matheson, V.A. (2011). Estimating Economic Impact using Expost Econometric Analysis: Cautionary tales, College of The Holy Cross, Department Of Economics Faculty Research Series, Paper No. 11-03.

11- Studenmund, A. H. (1997). Using Econometrics: A Practical Guide (3rd ed.). New York: Addison-Wesley.

12-T. B. Palaskas (1988). An Econometric Analysis of EEC Policy Ex-post and Ex-ante. An Application to Dried Vine Fruits in Greece. Journal of Oxford Development Studies. 17(1): 142-154.

13- USDA (2010). Grain: World Markets and Trade. Circular Series, FG 06-10, USA.

14-USDA-ERS (2008). Cost of Production Estimates. Economic Research Service. USDA Publications. Washington. DC.

15-USDA. (2008). Global Agricultural Supply and Demand: Factors Contributing to The Recent Increase in Food Commodity Prices. Economic Research Service, USDA Publications. Washington, DC.

16-World Bank (2008). Rising Food Prices: Policy Options and World Bank Response. Word Bank Publications, Washington. D.C.

3

Determinants of Repayment of Loan Beneficiaries of Micro Finance Institutions in Southeast States of Nigeria

Stephen Umamuefula Osuji Onyeagocha[1], Sunday Angus Nnachebe Dixie Chidebelu[2] and Eugene Chukwuemeka Okorji[2]

Abstract

The study investigated the loan repayment, its determinants and socio-economic characteristics of microfinance loan beneficiaries in the Southeast states of Nigeria. It was carried out in three states of the five southeast states. Using a multistage sampling technique, a total of 144 loan beneficiaries in the three segments of MFIs, namely; formal (commercial and development banks); semi-formal (NGOs-MFIs) and informal (ROSCAS, "Isusu" and co-operative societies) were randomly selected and interviewed in the three states. An ordinary least square (OLS) multiple regression analysis was carried out to isolate and examine the determinants of loan repayment from the respondents' perspective. Results showed that beneficiaries had low level of education, operated enterprises at a relatively small scale, had large family size and were of middle age. Further, it was found out that the majority of the respondents were involved in farming enterprise (crop and poultry) even though trading was the most prominent single non-farming enterprise (trading, processing and artisanship). The result affirmed that the informal sector respondents recorded the best repayment rate, followed by the respondents of semi-formal and the banks brought the rear. Outstanding among the determinants of loan repayments from the respondents' perspective were; loan size, level of education, experience, profitability and portfolio diversity. These, therefore deserve special attention in loan administration of MFIs.

Keywords:
Determinants of Repayment of MFIs Loan Beneficiaries, Nigeria

---

1 Department of Agricultural Economics, Federal University of Technology, Owerri, Imo State .
* Corresponding author's email: steveonyeagocha@yahoo.com

INTRODUCTION

Microfinance institutions are those institutions, which provide micro-credit, savings and other services to the productive poor. The focal point of many studies on microfinance dwells in the domain of poverty (Kanbur, 1987). Poverty is insufficiency of means relative to human needs. It is estimated that about 70% of Nigeria's population was poor and most of them live in rural areas and their major occupation is farming (CBN, 2002). Nigeria ranks as one of the 25 poorest countries in the world, having ranked 148 out of 173 countries surveyed.

Inadequate infra-structural facilities, poor social services, low technical education, unstable growth patterns of the economy and neglect of agriculture, among other factors are largely responsible for the despicable poverty situation in Nigeria. The fall in the quality of life of Nigerians to a reasonable extent is traceable to the neglect of the agricultural sector and the overdependence of the oil sector. The role of small-scale farming in economic development of developing countries such as Nigeria is inestimable. Apart from providing employment opportunities to about 80% of rural population, they supply food, fiber and raw materials for the populace, local industries and exporters. Production is characterized by small size of land (often less than one hectare) and use of crude implements, poor yielding seedlings, inefficient techniques, poor storage facilities, low level of education, to mention but a few. All these cumulated to poor income and resilient vicious circle of poverty. Similarly, micro-enterprises suffer from income anemia and vicious circle of poverty of the owners.

There is concern that poverty reduction strategy (PRS) to date have tended to emphasize the public provision of goods and services (roads, water, etc) and paid less attention to productive sectors (Cabral Lidia, 2006). To break these chains of poverty, ensure food security and industrial growth of developing nations, there is need for increase investment in the agricultural sector by both the government and the farmers. It therefore becomes imperative to expand and strengthen the financial institutions to play catalytic roles in this regard, especially in the area of providing machinery and tools, improved inputs and farmers' education. Several studies, including Feijo (2001) and Oyeyinka and Bo-lalarinwa (2009) have identified the positive impacts of credit in the operations of rural farmers.

Unfortunately, the formal financial institutions, especially the banks that are equipped to carry out these functions shy away from financing these farmers, on grounds that they are high risk ventures and involve huge administrative costs. This provided the opportunity for the informal financial sector such as money lenders (with its obnoxious interest rates), local co-operative societies, credit unions and thrift schemes that are less equipped to carry out this intermediation function, to key in and intensify credit delivery functions. Confirming this, the Central Bank of Nigeria (2005) noted that the formal financial system provides services to about 35% of the economically active population while the remaining 65% are excluded from access to financial services. According to the apex financial body, these 65% are often served by the informal sector through NGO-MFIs, friends, relations and credit unions.

Surprisingly, these informal institution apart from their high cost of credit, are performing exceedingly well in terms of loan repayment (which is the nightmare of formal financial institution). Also, their strong attribute is fast and efficient credit delivery with much less bureaucracies like collaterals which is replaced with trust and faith.

Loan repayment has been a critical problem of formal financial institutions in Nigeria. Studies in Imo State by Njoku and Odii (1991) recorded 27% repayment rate of the farmers, Njoku and Obasi (2001) in which 33.72% was recorded as repayment rate. This situation weakens the virility of the MFIs. According to CBN (2005), the weak capital base of the existing financial institutions, particularly the community banks (now transformed to micro finance banks), cannot adequately provide a cushion for the risk of lending to farmers and micro entrepreneurs without collateral. Further, poor repayment rate of credit reduces lenders net return thereby decreasing the ability of the institution to generate resources internally for institutional growth. In extreme cases, this may result in distress condition or outright liquidation of the institution. Besley (1994) affirmed that the issue of enforcing repayment constitutes a major problem in credit market. According to the author, enforcement problem arises in a situation in which the borrower is able but unwilling to repay the loan.

One way to tackle the loan repayment problem is to investigate the factors which affect the loan repayment of MFIs. Eze and Ibekwe (2007) in their study on determinants of loan repayment in Orlu Local Government of Imo State, Southeast, Nigeria, identified; loan size, age of beneficiaries, household size, and number of years of formal education and occupation as the key determinants. Similarly, Dayanandan and Weldeselassie (2008) in their study on loan determinants of small farmers in Northern Ethiopia, agreed with Eze and Ibekwe (2007) that amount of credit, educational status and occupation (nonfarm income) were potent factors in loan repayment. Other factors they isolated as potent were; experience, repayment period and ownership of livestock.

This study is aimed at providing answers to the hydra-headed repayment problem. It is reasonable to expect that an impressive loan repayment would be mutually beneficial to both the farmers/micro-entrepreneurs and the loan institutions. On the part of the farmers and micro entrepreneurs, good credit ratings would definitely attract more loans with which to procure improved inputs and implements. In such situation, efficiency would improve as well as profitability and these are capable of lifting them out of the vicious circle of poverty. For the financial institutions, which depend mainly on interest income for their institutional growth, prompt loan repayment would mean reduced cost and enhanced profitability and robust growth.

Therefore, the broad objective of this paper is to determine factors affecting repayment rate of loan beneficiaries of MFIs in the Southeast States of Nigeria. The study specifically investigated the social-economic characteristics of the respondents; determine their loan repayment rate and its determinants.

MATERIALS AND METHODS

The study was carried out in Southeastern states of Nigeria comprising of Abia, Anambra, Ebonyi Enugu and Imo States. The area had a population of 25.9 million, which is about 30% of the national population (2006). The Southeast states are among the mostly densely settled area of the country, with average population density of 247 persons per square kilometer as against the national average of 96 persons per square Kilometer (NPC, 2006).

The choice of the area was because of intense activities of self help groups in various economic activities, including agriculture in the area. Also, there is a high degree of socio-cultural homogeneity in the study area as the inhabitants are mainly Igbos, known mainly for their hard work, self-reliance and economic

prowess.

Multi-stage sampling technique was employed in the selection of

respondents who were mainly loan beneficiaries of commercial, development, community (micro finance) banks, NGO-MFIs groups and the local Isusu, co-operatives, ROSCAS members. The sample frame was provided by the Central Bank of Nigeria for NGO – MFIs; the banks and the local extension agents of the local government council.

In stage one, three out of five south-east states were purposively selected based on intensity of MFIs activities.

Stage two involved the selection of MFIs, which were stratified into formal, semi-formal and informal. From each stratum, four institutions were selected randomly. Thus, giving a total of 12 MFIs per state and 36 MFIs for the three states selected.

Finally, from each of the 12 MFIs in a state, four respondents were selected, randomly. Thus, giving a total of 48 respondents per state, and 144 respondents for the three states, representing the south-east states. The respondents were selected from 28 out of 57 LGAs of the three states and this represented about 49% coverage of the total number of the LGAs. The 28 LGAs came into the sample by chance factor as no deliberate effort was made to choose them.

From the selected respondents, which involved five enterprise-types namely; crop and poultry farmers, traders, agro-processors and artisans; calibrated as farming and non-farming activities. Primary data were collected with the aid of a structured and pre-tested questionnaire. The secondary data were collected from journals, textbooks, annual accounts, return from banks, UNDP and CGAP (the consultative group to assist the poorest) websites.

The data collected were subjected to both descriptive and quantitative techniques, to realize the objectives of the study. The OLS multiple regression analysis was used to determine factors which affected repayment rate of loan beneficiaries. The linear functional form was adjudged the most appropriate for a repayment function. The model is stated as follows:

$Y = f(X_1, X_2, X_3 --- X_{13}, e)$

$Y_1 =$ Repayment rate (%)

$X_1 =$ Loan size (N)

$X_2 =$ Dependency ratio (children as percentage of total households size)

$X_3 =$ Level of education (year of formal education)

$X_4 =$ Age (years)

$X_5 =$ Enterprise type (dummy variable: farming enterprise =o, and non farming enterprise = 1)

$X_6 =$ Experience (years)

$X_7 =$ Profitability of respondents enterprises (N)

$X_8 =$ Training (total no. of days per year)

$X_9 =$ Interest rate (%)

$X_{10} =$ Repeat loan (%)

$X_{11} =$ Gender factor (percentage of group members who are female)

$X_{12} =$ Shocks (No. of family emergencies, crop/income loss due to

incidence of pests and diseases, major social

events that occurred in the previous 18 months)

$X_{13} =$ Portfolio Diversity (proportion of members that have secondary occupation).

e = error term.

RESULTS AND DISCUSSION
Socio-economic characteristics

The socio-economic characteristics are important sign posts in explaining the behaviour of the farmers and micro entrepreneurs in certain actions such as management and loan repayment decisions. They complement the results of the technical or quantitative analysis such as OLS multiple regression. Some of these characteristics are summarized in the tables.

Table 1(a) is the distribution of the mean value of some economic indices of the respondents. The majority of the respondents (63.2%)

Table 1(a): Distribution of the mean values of some economic indices of the respondents

Socio-economic Characteristics	Mean Value
Majority of sex: female	63
Marital Status	80
Family size	10
Age (years)	41
Experience (years)	8.8

Table 1(b): Distribution of Number of Years Spent in School

No of Years Spent in School	Frequency	%
None	43	29.9
1 – 6	31	21.5
7 – 12	46	31.9
13 -16	24	16.7
Total	144	100.0

Table 1(c): Distribution of Primary Occupation

Occupation	Frequency	%
Trading	48	33.3
Crop farming	41	28.5
Agro – processing	17	11.8
Poultry farming	27	18.8
Poultry farming	11	7.6
Total	144	100.0

Table 1(d): Distribution of respondents by enterprises size (turnover: Naira for traders, agro processors and artisans)

Class	Frequency	%
Less than 20,000	1	1.32
21,000 -51,000	2	2.64
52,000 – 82,000	2	2.64
83,000 – 113,000	10	13.15
114,000 – 113,000	10	13.15
114,000 – 144,000	21	27.63
145,000 – 175,000	18	23.68
176,000 – 206,000	12	15.79
Greater than 206,000	10	13.15
Total	76	100.0

Table 1(e): Distribution of poultry farmers by enterprises size (stock of birds)

Class	Frequency	%
less than 50	1	3.7
51-101	15	55.6
102 -152	8	29.6
152-203	2	7.4
Greater than 203	1	3.7
Total	27	100.0

Table 1(f): Distribution of crop farmers by farm size (hectare)

Class	Frequency	%
01 or less	12	29.27
0.2 -0.4	18	43.90
0.5-0.7	7	17.07
0.8-1.0	3	7.32
Greater than 1.0	1	2.44
Total	41	100.0

were female and male constituted only 36.8 percent. Eighty percent of the respondents were married and by implication were likely to have families, while 20% were single. On age, about 55% of the respondents were of middle age bracket and above, with about 45% being youths. The respondents have relatively large family with 10 as mean family size as against the recommended national figure of six. Over 70% of the respondents had eight years and above in experience in work with a mean figure of 8.8 years.

Table I (b) is the distribution of the respondents by level of education. It showed that about 70% of the respondents are literate and about 30% were not literate. This suggested that education was still a problem. Literacy level impacts positively in productivity and efficiency of farmers through adoption of technology and innovations.

Table 1(c) is the distribution of respondents by primary occupation. It suggested that trading was the primary occupation of the greatest number (33%) of the respondents. However, on the aggregate, farming constituted about 60% of the respondents' primary occupation while non-farm enterprise constituted about 40%. About 40% of the non-farming respondents have farming as secondary occupation.

Tables 1(d), (e) and (f) are the distribution of respondents by enterprise size (Naira) for traders/processors/artisans; stock of birds for poultry farmers and farm size (hectares) for crop farmers, respectively. Table 1(d) showed that over 71% of the respondents had a turnover of less than N144, 000 per annum. This suggested that the respondents were of low income group. Table 1(e) showed that over 80% of the poultry farmers had not more than 152 birds in their

Table 2: Distribution of Respondents by Sources of Loans

Class	Frequency	%
Co-op Soc.	17	11.8
NGO/MFIs	53	36.8
Commercial Banks	15	10.4
ROSCAS	24	16.7
DFI (NACRDB)	21	14.6
Community Banks (MFBs)	14	9.7
Total	**144**	**100.0**

stock. The mean stock of birds for these farmers was 102 birds suggesting small-scale operations. Table 1(f) showed that over 90% of the crop farmers owned or cultivated not more than 0.7 hectares of land. The mean size of farm of the respondents was 0.46 hectare, suggesting that they were operating mainly on a small-scale.

Sources of Loan and Repayment Rate

Table 2, showed the distribution of respondents according to sources of their loan. The NGO-Microfinance Institutions provided loan to 38.8% of the respondents. This was followed by Rotation Savings and Credit Association (ROSCAS) (16.7%), NACRDB (14.6%), co-operative societies (11.8%), commercial banks (10.4%) and Community Banks (or Microfinance Banks) (9.7%).

On loan repayment, this was segmented into prompt repayment (for those repayments that were effected as scheduled) and overall repayment (for those repayments that were effected not as scheduled and of course, which involved recovery costs on the part of the financial institutions) as indicated on Table 3(a). On prompt repayment, the respondents of informal institutions recorded an average of 90%, repayment rate followed by semi-formal institutions (NGO-MFIs) 73.57%

Table 3: Loan Repayment of Respondents

Enterprise (or MFI Category)	Frequency	Repayment(%)	
		Prompt	Overall
a) MFI Categorization			
Formal	12	43.04	56.58
Semi- Formal	12	73.57	84.91
Informal	12	90.00	100.00
b) Enterprise Type			
Crop Farming	41	55.47	
Poultry Farming	27	41.20	48.33 (AV.)
Trading	48	78.78	
Agro-processing	17	70.79	
Artisans	11	61.50	70.35 (AV.)

Table 4: Distribution of Respondents by reasons for Default

Item	Frequency	%
Item	17	11.8
Poor harvest due to crop failure	10	7.0
Low market price	41	28.5
Incidence of Pest and Diseases	24	16.3
Untimely loan disbursement	19	13.2
Family commitments	50	35.0
	144	**100.0**

and formal institutions (banks) 43.04%. On overall repayment, the respondents of informal financial institutions recorded 100% repayment rate. This was followed by semi-formal institutions 84.91% and formal institutions 56.58%. Table 3(b) indicated that the respondents in trading repaid about 79% of their loan promptly. This was followed by agro-processors (about 71%). In general, non-farming enterprises on the average repaid about 70% of their loan as against 48% of farming enterprises. This could be attributed to the complex and risky nature of farm enterprises.

Table 4 showed the reasons for default. It showed that family commitments ranked highest (35%) among the reasons adduced for default. This was followed by low market prices (28.5%), incidence of pests and diseases (16.3%), untimely disbursements (13.2%) and crop failure (7%). Family commitments (like school fees, extended family problems, burial and other cultural ceremonies) were a big burden on the respondents

as well as low market prices, especially during harvest, occasioned mainly from lack of poor storage facilities.

Determinants of Loan Repayment of Respondents

Table 5 showed the factors, which affected loan repayment and were calibrated as determinants of loan repayment. It indicated that out of 13 explanatory variables, five were the most potent factors. The Coefficient of Multiple Determination (R^2) was 0.5022, suggesting that about 50% in the variation of loan repayment was accounted for by the variations of the explanatory variables. This suggests that there may be other factors not included in the model. If $R^2 = 1$, it implies that there was 100% explanation of the variation in loan repayment by the explanatory variables or regessand. However, if $R^2 = 0$, it means that the explanatory variables do not explain any changes in the criterion variable or loan repayment. The F-value is used to

Table 5: Determinants of Loan Repayments of Respondents:

VARIABLE	UNIT	COEFFICIENT	T-RATIO
Loan Size	Naira	12.0318	2.9272*
Dependency ratio	Percent	-7.1043	-1.1422
Level of education	Years	15.9122	2.6372*
Age	Year	-6.0359	-1.0751
Enterprise type	Dummy	8.2134	1.0359
Experience	Years	10.4494	3.3368*
Profitability Index	Number	17.0318	4.0632*
Training period	Days	9.4227	1.1725
Interest rate	Percent	-5.0389	-1.2260
Repeat loan	Dummy	9.1163	1.1339
Gender factor	Percent	11.0295	1.0870
Shocks	Likert Ranking	-15.0214	-1.0019
Portfolio diversity	Dummy	6.9943	3.3928*
Constant	39.9133		
R2	0.5022		
F-value	10.0884		
N	144		
d.f.	130		

LOS = *5%

test whether or not there is significant impact between the dependent variable and the independent variables. In regression equation, if F-calculated is greater than F-tabulated, then there is significant impact between the dependent variable and the independent variables. If otherwise, the reverse is the case.

The five potent variables which affected loan repayment were; loan size, level of education, experience, profitability and portfolio diversity and they are subsequently discussed.

(a) Loan Size

Loan size was significant at the 5% LOS and was positively related to repayment rate. This implies, the greater the size of the loan, the lower the default. This was true up to a certain point as there was an optimum amount of loan (or funds) that would be required to break even in projects. Moreover, it is contended that bigger loans make possible larger investments with potentially higher returns. About 75% of the loan beneficiaries indicated that the sizes of their loans were inadequate, thus supporting this viewpoint. Also, Njoku and Obasi (2001) isolated loan size, among two other variables, that are important and have positive relationship with loan repayment under ACGFS in Imo State.

Similarly, Olagunju (2007) in his study on the impact of credit use, agreed with this view point.

The second perspective to this variable was the larger the loan, the higher is the borrower's cost of delaying payment. A larger loan is more difficult to repay if allowed to accumulate especially where there are compounding interest and sanctions. This second factor puts pressure on the borrower to reduce late payments and serious default. In the sample, recorded incremental penalty rate of interest for delay payment was minimal.

(b) Level of Education

The level of education was significant at the 5% level, and was

positively signed as hypothesized. This suggests that as the level of education improved the beneficiary also improved the ability to read and write and in the process, improved dexterity in the occupation, which concomitantly improved profit and the capacity to repay loans. This is in agreement with Coelli and Battese (1996) in

India.

respondents were 6.4 while the figure for non literate respondents was 30%, which suggested that there were lots of room for improvement in their education status.

(c) Experience

The coefficient of experience was positive and significant at 5% level suggesting that the length of experience in occupation was a potent factor in loan repayment. This was because experience provided the compass with which the entrepreneur navigated the turmoil business environment and was a veritable decision tool. The result and that of Parikh and Mirkalan (1995) supported this hypothesis. The respondents had eight or more years in terms of business experience. However, Ogundare (2009) reported a negative coefficient of age and farming experience, which implies that output decreased as each of these variables increased. It suggests that the more the years of experience of the farmer and by implication, the older in age and the less productive and the tendency of increasing risk aversion.

(d) Profitability

The coefficient of profitability index was positive and significant at 5% level and was in consonance with hypothesis, which stated that profitability index (ratio of income to costs) had direct and strong relationship with repayment. This was because difficulties in repayment arose whenever a business in unprofitable. It is an indication or index of management ability. In the event of not making profit, enterprises including NGOs (which are expected to break-even), become unsustainable.

(d) Portfolio Diversity

This indicates the proportion of beneficiaries who have secondary occupation. It is therefore an indicator of asset portfolio diversity within the group/respondents. The study showed that the majority (66%) of the respondents have trading as their secondary occupation. Due to diversity, income within groups tended to be less covariant, thus making it easier to bail out errant members. As hypothesized, the coefficient of the variable was positively signed and significant at 5% level, indicating strong relationship.

The linear equation can generally be represented thus:

$Y1 = 39.9133 + 12.0318X1 - 7.1043X2 + 15.9122X3 - 6.0359X4 +$

(2.93*)   (2.64*)

$8.2134X5 + 10.44946 + 17.0318X7 + 9.4227X8 - 5.0389X9$

(3.34*)   (4.06*)

$+9.1163X10 + 11.295X11 - 15.0214X12 + 6.9943X13 + 6.0038$

(3.39)*

$R^2 = 0.5022$ F-Value = 10.0884   *1%LOS

CONCLUSION

The respondents are certainly micro/small scale operators with low income, poor educational background and relatively large family size and its attendant burden and challenges. The respondents were of middle age and females were predominant. Farming was the main occupation and trading constituted a third of the respondents' occupation. Nevertheless, half of the trading respondents have farming as their secondary occupation.

The major source of their loans were the informal sector namely; NGOs-MFIs and ROSCAS. The respondents of the informal sector performed most creditably in terms of loan repayments. This was followed by the semi-formal (NGOs-MFIs) and the banks brought the rear. This perhaps may be due to the fact that screening, monitoring and enforcement of payment were carried out by the group members themselves. In terms of enterprise type, trading was found to be the most important with respect to loan repayment. This was followed by agro-processors and artisan (others). Crop and poultry farming brought the rear. In general, non-farming enterprises performed better than farming enterprises in terms of loan repayment. The difference could be attributed to the complex and risky nature of farming, hence the need for extra ordinary support for farming enterprises.

In terms of loan administration and repayment, adequate attention should be paid to loan size, level of education, experience, profitability, portfolio diversity. These constituted the determinants of loan repayment from the respondents' perspective and therefore deserve more focus and attention.

Further, formation of autonomous cooperative societies, provision of storage facilities and reduction of some associated expenses that affect family commitments (e.g. school fees) will help reduce loan default.

REFERENCES

1- Besley, T. (1994). How Do Market Failures Justify Interventions in Rural Credit Markets? World Bank Research Observer. 9(1): 27-48.

2- Central Bank of Nigeria (2002). Baseline Survey of Micro Finance Institutions in Nigeria. CBN Abuja. P2.

3- Central Bank of Nigeria (2005). Micro Finance Policy, Regulatory and Supervisory Framework for Nigeria. CBN Abuja P. 2

4- Coelli T and G.B. Battese (1996). Identification of Factors which Influence the Technical Inefficiency of Indian Farmers. Australian Journal of Agricultural Economics. 40: 103-128.

5- Dayanandan, R., & Weldeseassie, H. (2008). Determinants of Loan Repayment Performance among Small Farmers in Northern Ethiopia. Journal of African Development Studies. (Abstract).

6- Eze, C.C., & Ibekwe, U.C. (2007). Determinants of Loan Repayment under Indigenous Financial System in Southeast, Nigeria, Medwell Journals. 2(2): 116-120.

7- Feijo, R.L.C. (2001). The Impact of a Family Farming Credit Program on the Rural Economy of Brazil. www.anpec.org.br/encontro 2001.

8- Kanbur, R.S.M. (1987). Measurement and Alleviation of Poverty, with an Explanation to the Effect of Microeconomic Adjustments. Cabral Lidia (2006). Poverty Reduction Strategies, and the Rural Productive Sectors: What Have We Learnt, What Else do We need to ask? ODI Natural Resources Perspective 100, UK.

9- Njoku, J.E. and M.R. Odii (1991). Determinants of Loan Repayment under the Special Emergency Loan Scheme (SEALS) in Nigeria: A Case Study of Imo State. African Review of Money, Finance and Banking. 1: 39-52.

10- Njoku, J.E. & Obasi, P.C. (2001). Loan Repayment and its Determinants under the ACGS in Imo State, Nigeria, Africa Review of Money Finance and Banking, (Abstract).

11- Ogundare, K. (2009). Technical Efficiency of Farmers under Different Multiple Cropping Systems in Nigeria. Journal of Tropical and Sub-Tropical Agro Ecosystem 10: 117-120.

12- Ohajianya, D.O., & Onyenweaku, C.E. (2003). Demand for Community Bank Credit by small Holder Farmers in Imo State, Nigeria. Journal of Sustainable Tropical Agricultural Research. (Abstract).

13- Olagunju, F.I. (2007). Impact of Credit Use on Resource Productivity of Sweet Potato Farmers in Osun State Nigeria. Journal of Social Sciences 14 (2): 175-178.

14- Oyeyinka, R.A., & Bolarinwa, K.K. (2009).Using Nigerian Agricultural and Cooperative Bank, Small Holder Loan Scheme to increase agricultural production in rural Oyo State, Nigeria". International Journal of Agricultural Economics and Rural Development. (Abstract).

15- Parikh, A.E., & Mirkalan, S. (1995). Measurement of Economic Efficiency in Pakistan, American Journal of Agriculture. 32:53-61.

16- National Population Commission (2006). Population Census of Federal Republic of Nigeria, Analytical Report of the National Level, Abuja, NPC.

4

Analysis of Technical Efficiency of Smallholder Cocoa Farmers in Cross River State, Nigeria

Agom, Damian Ila[1], Susan Ben Ohen[1], Kingsley Okoi Itam[1] and Nyambi N. Inyang[2]

Abstract

The technical efficiency involved in cocoa production in Cross River State was estimated using the stochastic frontier production function analysis. The effects of some selected socio-economic characteristics of the farmers on the efficiency indices were also estimated. The study relied upon primary data generated from interviewing cocoa farmers using a set of structured questionnaire. A multi-staged random sampling technique was adopted in selecting two hundred (200) cocoa farmers from Ikom Agricultural Zone in the state. The data on the socio-economic characteristics of the farmers were analyzed using descriptive statistics, while the stochastic production function, using the Maximum Likelihood Estimating (MLE) techniques was used in estimating the farmer's technical efficiency and their determinants. Result of the analysis showed that farmers were experiencing decreasing but positive returns to scale in the use of the farm resources. The efficiency level ranged between 0.20 and 0.93 with a mean of 0.69. The result of the generalized Likelihood Ratio (LR) tests confirmed that the cocoa farmers in the area were technically inefficient. The major contributing factors to efficiency were age of farmers, farm size, level of education, sex of farmer and age of the farms. The study observed that there is enough room to improve efficiency with the farmers' current resource base and available technology and concluded that policies that would directly affect these identified variables should be pursued.

Keywords:
Cocoa production, Technical efficiency, Stochastic frontier, Likelihood ratio, Maximum likelihood ratio

[1] *Deptartment of Agric Economics, University of Calabar, Calabar, Nigeria.*
[2] *Commercial Agric Development Project (CADP), Cross River State, Nigeria.*
* *Corresponding author's email: agomd@yahoo.com*

INTRODUCTION

Agriculture in developing countries is characterized by low productivity leading to low farm incomes. In Nigeria, cocoa production is characterized by several problems that lead to low productivity. This has resulted to a fall in percentage share of cocoa output. As Amos (2007) notes, two reasons are said to be responsible for the fall in percentage share of cocoa output. First is the negligence of the agricultural sector by the past administration due to the discovery of the petroleum resources that now accounts for the bulk of foreign exchange earnings. Second is the endemic problem in the cocoa industry. Therefore, increasing productivity will increase the percentage share of cocoa production. Analysis reveals that increasing agricultural production has probably been the simple most important factor in determining the speed and extent of poverty reduction. Most of these evidences are derivable from the Green Revolution in Asia (Adeniran, 2007). In China rapid productivity gains achieved largely through technological advances of the Green Revolution directly increased producers income and labourers' wages by lowering the price of food and by generating new livelihood opportunities as success in agriculture provides the basis for economic diversification. The importance of productivity is that it gives a measure for efficiency.

In Nigeria, there have been studies on farm level efficiency in tree crop production and very few have focused on cocoa production. Among these are studies by Giroh et al., (2008) who carried out analysis of the technical inefficiency of gum Arabic based cropping patterns among farmers in the gum Arabic belt of Nigeria and that of Amos (2007) whose study is analysis of productivity and technical efficiency of small holder cocoa farmers in Nigeria. The authors employed the stochastic frontier production function analysis in their studies. However we do not have such studies in Cross River State which is a major cocoa producing area in Nigeria. Recent studies carried out on cocoa production in Cross River State examined the Socio-economic variables and cocoa production (Oluyole and Sanusi, 2009). Fertilizer use and cocoa production and Investment in cocoa production in Nigeria: .A Cost and Return analysis of three cocoa production management system in Cross River State cocoa belt by Nkang et al., (2009). None of these studies examined the technical efficiency of the cocoa farmers.

Objectives:

This study was carried out to provide estimates of levels of technical efficiency of cocoa farmers in Cross River State using farmers in Ikom Agricultural zone, where there is a high concentration of cocoa farmers in the state. The study was interested in whether the cocoa farmers were fully technically efficient, the current level of efficiency and factors that influenced efficiency. The study therefore was to estimate the level of and determinants of technical efficiency among cocoa farmers in Cross River State.

MATERIALS AND METHODES
Study area

This study was conducted in the two major Cocoa Producing Local Government Areas in Cross River State; Etung and Ikom. Cross River State is located in the Niger Delta region of Nigeria. It is bounded in the North by Benue State, in the South by Atlantic Ocean, in the East by Cameroon Republic and on the West by Akwa Ibom State, Abia and Ebonyi States. The state lies within latitude 40° 4, South and 60°30, North and between longitude 8° and 9° 00" East of the equator. It has three distinct ecological zones, the mangrove forest to the south, the tropical rainforest in the middle and the guinea savanna to the north. The annual mean rainfall ranges between 1500mm and 2000 mm.

Sampling procedure and sample size

The multistage random sampling technique was adopted for this study. The first stage involved a purposive selection of two (2) Local Government Areas in the Ikom Agricultural Zone- Ikom and Etung. This is because Ikom and Etung are the major cocoa producing areas in Cross River State. The second stage involved the random selection of Five (5) villages from each of the selected Local Government Areas, giving a total of Ten (10) villages. The villages selected in Ikom were Akparabong, Ayukasa, Okondi, Alok and Ajassor, while those selected in Etung were Effraya, Agbokim, Bendeghe Ekim, Abijang and Abia. A simple random selection of twenty (20) farmers from each of the selected villages was carried out making up a total of one hundred (100) farmers from each of the two Local Government Areas which gave us 200 cocoa farmers for the study. Information was obtained on socio-economic characteristics of the farmers, output, labour, farm size and prices of variables using a set of structured questionnaire.

Model specification

Descriptive statistics including mean, standard deviation and variances were used to analyse the socio-economic characteristics of farmers while the stochastic production function was used to analyze the level of technical efficiency. The production technology of the cocoa farmers was assumed to be specified by the Cobb-Douglas frontier production function (Tadesse, and Krishnamorthy, 1997; Amos, 2007).

The specified cocoa production function was given as follows;

$$In\ Y = In\ \beta_0 + \beta_1 In X_{1i} + \beta_2 In X_{2i} + \beta_3 In X_{3i}\ \beta_4 In X_{4i} + V_i - U_i \qquad (1)$$

Where;

Y = Quantity of cocoa produced (kg)

X_1 = Farm size (hectares)

X_2 = Quantity of fertilizer (kg)

X_3 = Quantity of fungicide (litres)

X_4 = Labour (man days)

B_0 = Y - intercept

B_1 to β_4 are coefficients to be estimated and V_i and U_i are error terms. It is expected that β_1, β_2, β_3, and β_4 will have positive signs.

Determinants of technical efficiency

The influence of some socio-economic factors on the computed technical efficiency was determined by incorporating the socio-economic factors directly in the frontier model, because they have influence on efficiency (Kalirajan, 1981). The technical efficiency model was specified as:

$$U_i = a_0 + a_1 Z_{1i} + a_2 Z_{2i} + a_3 Z_{3i} + a_4 Z_{4i} +$$

$$a_5 Z_{5i} + a_6 Z_{6i} + a_7 Z_{7i} \qquad (2)$$

Where

U_i = Technical efficiency

Z_1 = Sex of farmer (dummy)

Z_2 = Marital status (dummy)

Z_3 = Age of farmer (years)

Z_4 = Education (years spent in school)

Z_5 = Family size

Z_6 = Age of farm (years)

Z_7 = Farm size (ha)

a_0 = y - intercept

a_1 to a_7 are coefficients that were estimated. It was expected that a_3 would have a negative sign, while a_2, a_4, a_5, a_6 and a_7 would have positive signs. The sign of a_1 was indeterminate.

In determining the level of technical efficiency of the cocoa farmers and analyzing the determinants of technical efficiency among the cocoa farmers, a generalized likelihood ratio (LR) test was used to test the hypothesis of full technical efficiency effects defined as

$$LR = -2\ In\ (logH_1 - logH_2) \qquad (3)$$

Where, H1 is the log – likelihood function of the average function. H2 is the log- likelihood function of the frontier function. The value has a mixed chi-square distribution with degrees of freedom equal to the number of parameters plus one. A computer programme frontier version 4.1 by Coelli (1994) was used in the computation, while the testing of the parameters was done at 1 and 5 percent levels of significance.

RESULTS AND DISCUSSION

Socio-economic characteristics of the sampled cocoa farmers

The distribution of sampled cocoa farmers according to their sex, age, marital status, educational level, family size and farming experience is presented in Table 1. The results indicates that majority (88.5%) of the farmers were males while few (11.5%) were females. The cultural setting of the area allows the males to have easy access to land especially, where majority of them are the heads of their respective households.

The table also indicates that majority of the farmers (44.5%) were within the 47 to 57 years age bracket. This was closely followed by the farmers with age 36-46 years (32%). Farmers that were in the minority constituted 1.5% and these farmers were above the 68 years age bracket. This shows that about 86% of the farmers were in their most economically active age bracket (25-57) years. However, there was a widespread of farmers among all the age groups, implying that cocoa farming was embraced by all the age groups. The results also showed that most of the farmers (68.5%) were married while 31.5% were single (Table 1).

Furthermore, the distribution of the cocoa framers according to their educational level (number of years spent in school), shows that majority (76.5%) of the farmers had attained one level of formal education or the other. The mean level of educational attainment (years spent in school) of the farmers in the area was about 6 years, with 11% of the farmers having tertiary education. This is an indication that some graduates were involved in cocoa farming in the study area. This is a good pointer to improved productivity as the level of education is a tool with which an individual could be efficient at whatever endeavour being undertaken by the individual (Oluyole and Usman, 2006). As regards family size, a high proportion (86.5%) of the farmers had family sizes of 5 persons and above, while 13.5% had less than 5 persons in their household. The mean family size of the cocoa farmers was 7 persons. Effiong (2005) reported that a relatively large household size enhances the availability of family labour which reduces constraints on labour cost in agricultural production.

The table also shows that a very high proportion (99%) of the farmers had between 5 and above thirty (30) years of experience in cocoa farming. The mean farming experience was about 15 years. Farmers sometimes count more on their experience than educational attainment in order to increase their productivity (Nwaru, 2004). The result implies that a good number of the farmers are experienced farmers and therefore are expected to have higher technical efficiencies.

Mean output and other production variables in cocoa production in Cross River State

The statistics of the production variables obtained from cocoa farmers in the study area are

Table 1: Distribution of socio-economic characteristics of sampled cocoa farmers

Variables	Frequency	Percentage
Sex:		
Male	177	88.5
Female	23	11.5
Total	200	100
Age of farmer (years):		
25 – 35	19	9.5
36 – 46	64	32.0
47 – 57	89	44.5
58 – 68	25	12.5
>68	3	1.5
Total	200	100
Means	47.70 (8.91)	
Marital status:		
Single	63	31.5
Married	137	68.5
Total	200	100
Mean	2.02(1.61)	
Educational level (school years):		
0	47	23.5
6	81	40.5
8	14	7.0
12	36	18.0
14	14	7.0
16	8	40
Total	200	100
Mean	6.45 (4.49)	
Family size:		
<5	27	13.5
5 – 7	76	38
8 – 10	54	27
11 – 13	24	12
14 – 16	9	4.5
17 – 19	7	3.5
>19	3	1.5
Total	200	100
Mean	7. 11 (4.66)	
Farming experience (years)		
<5	2	1.0
5 – 10	74	3.7
11 – 15	54	2.7
16 – 20	33	16.5
21 – 25	19	9.5
26 - 30	6	3.0
>30	12	6.0
Total	200	100
Mean	15.22 (7.70)	

Source: Field Survey 2010
Note: Values in parentheses are standard deviations.

summarized in table 2. The mean output of cocoa farmers in the area was 2428.10kg/annum/farmer. This is relatively high compared to figures of less than two tonnes recorded elsewhere. This may be related to the age of the farms as most farms in Cross River State are within the productive age of 11 to 40 years (Table 3) compared to farms in other parts of Nigeria. For labour, the

Table 2: Summary statistics of output and other variables for sampled Cocoa farmers

Variables	Minimum	Maximum	Mean	Standard Deviation
Output (kg)	760.00	76200.00	2428.10	2343.53
Farm size (ha)	1.00	40.00	6.90	6.35
Fertilizer (kg)	0.00	200.00	18.50	37.09
Fungicide (litres)	0.00	2500.00	1142.80	644.37
Labour (man days)	18.00	93.00	51.49	14.65
Age of farmer (years)	25.00	71.00	47.70	8.91
Family size	1.00	26.00	7.00	4.66
Farming experience (years)	4.00	50.00	15.22	7.70

Source: Derived from Field Survey Data 2010

mean man- days used by the farmer during the production season were 51.50 man days. Farm sizes in Nigeria have been described as small, medium or large scale, if they fall into categories of less than 5ha, between 5ha and 10ha, or more than 10ha, respectively (Upton,1972). Most of the farms in Nigeria are of small to medium scale categories. The average farm size among the cocoa farmers in the study area is 6.90 hectares scattered in different locations in the locality, hence the small holdings. It was observed that majority of the cocoa farmers in Cross River State did not use fertilizer in cocoa production. The mean fertilizer used by the farmers was 18.50, which is very low. The result is in line with Oluyole and Sanusi, (2009) which reported that 98.13% of cocoa farmers in Cross River State did not use fertilizer in cocoa production.

An average of 1142.80 litres of fungicide was applied by the cocoa farmers (Table 2). The high rate of application may be due to the less resistance of the cocoa variety to infection.

Classification of cocoa farmers according to age groups and their mean output in the study area

The sampled farms were grouped according to their age groups and mean output and majority (92%) of the farms are within the productive age of 11 to 40years age group (Table 3). The mean output initially increased within this age as the cocoa trees get fully established, and thereafter output declines as shown in the table.

Maximum likelihood estimates of stochastic production frontier function for cocoa farmers in Cross River State

The coefficient of the Maximum Likelihood Estimates (MLE) of the parameters of the stochastic frontier models of cocoa farmers is shown in Table 4. The variance parameters of the stochastic frontier production function are represented by sigma squared (δ^2) and gamma (y). From the table, the sigma squared is 0.171 and significantly different from zero at one percent level. This indicated a good fit and correctness of the distributional form assumed for the composite error term. Gamma (y) indicates that the systematic influences that are unexplained by the production function are the dominant sources of random error. The gamma estimate which is 0.327 and significant at five percent level shows the amount of variation resulting from the technical efficiencies of cocoa farmers

Table 3: Age group of farmers and mean output

Age group	Frequency	Percentage	Mean output	Standard deviation
5 - 10	5	2.5	6305.2	6305.2
11- 20	81	40.5	8064.1	8064.1
21-30	81	40.5	8797.0	8797.0
31- 40	22	11.0	10448.0	10448.0
41- 50	8	4.0	7465.1	7465.1
>50	3	1.5	6573.0	6573.0
	200	**100**	**(2428.1)**	**(2428.1)**

Source: Field Survey Data 2010
Note: The values in parenthesis are the mean output and standard deviation of the farms respectively and do not represent the total mean output and standard deviation of the groups.

Table 4: Maximum likelihood estimates of the stochastic production function for cocoa production

Variable	Parameters	Coefficients	Standard errors	Standard errors
Constant	β_0	7.450	0.141	52.837***
Farm size	β_1	0.155	0.016	9.688***
Quantity of fertilizer	β_2	0.013	0.001	1.30
Quantity of fungicide	β_3	0.002	0.004	0.50
Labour	β_4	0.009	0.003	3.00***
Diagnostic statistics				
Gmma (Y)	Y	0.327	0.144	2.71**
Sigma square	δ_2	0.171	0.037	4.622***
Log likelihood function		-88.08		
Likelihood ratio (LR)	λ	71.42		

Source: Computed from Field Survey Data 2010 using Frontier 4.1
Note: *** P < 0/01, ** P < 0/05

in the study area. This means that more than 32% of the variation in farmer's output is due to difference in technical efficiency.

The result further shows that the signs of all the estimated coefficients of the stochastic production frontier are positive which is consistent with a priori expectation. This implies that there is a positive relationship between the level of output of cocoa and farm size, the quantity of fertilizer, fungicide and labour used. This is expected as the level of production depends largely on the quantities of these inputs used on the farm. This can only be up to a level that is considered optimal after which farmers will be operating at sub optimal level. However, the coefficients of the slope, farm size and labour were significant at one percent indicating that farm size and labour are important determinants of cocoa output.

Determinants of technical efficiency in cocoa production

The analysis of the efficiency model shows that the signs of the estimated coefficients in the efficiency model have important implications on the technical efficiency of cocoa farmers in the study area. The coefficient of sex ($z \leftarrow \leftarrow 1$) had a positive sign indicating that the male farmers obtain higher levels of technical efficiency than their female counterparts in the area. Cocoa farming is dominated by males in the area. This is so because cocoa farming is a tedious job and requires more strength which females may not be able to provide (Oluyole and Sanusi, 2009).

The coefficient of marital status (Z_2) is indeterminate. However, it is negative as shown in Table 5.

The coefficient of age (z_3) is also positive. This does not agree with a prior expectation. As farmers age, there is a tendency that productivity will continue to fall owing to their declining strength. However, this result could be attributed to the fact that most of the farmers in the study area started farming at early age. Hence, the older they are the more experienced and efficient they would be, since farmers' experience increases with the number of years spent in farming, it implies that the longer the time spent in

Table 5: Determinants of Technical Efficiency

Variable	Parameter	Coefficient	Standard error	t-ratios
Constant	a_0	-1.278	0.699	-1.828
Sex of farmer	a_1	0.009	0.415	0.022
Marital status	a_2	-0.209	0.135	-1.548
Age of farmer	a_3	0.011	0.136	0.801
Educational level (sch yrs)	a_4	0.054	0.204	0.265
Family size	a_5	-0476	0.236	-2.016**
Age of farm	a_6	-0.012	0.723	-0.017
Farm size	a_7	0.081	0.010	8.10***

Source: Computed from Field Survey Data 2010 using Frontier 4.1
Note: *** P < 0/01, ** P < 0/05

Table 6: Elasticity of production and returns to scale

Variables	Elasticity
Farm size	0.155
Quantity of fertilizer	0.086
Quantity of fungicide	0.420
Labour	0.094
RTS	0.179

Source: Field Survey, 2010

farming the more experience they are.

Furthermore, the coefficient of educational level (z_4) was positive but not significant. This implies that the level of technical efficiency of the farmer increases with the level of education but not significantly. However, the result agrees with a prior that technical efficiency should increase with increases in years of schooling of the farmers since education and adoption of innovation were expected to be positively correlated.

The coefficient of family size (z_5) has a negative sign and is significant at the five percent level. This implies that increase in family size will lead to decrease in technical efficiency. This result does not agree with a prior expectation. However, given that the cocoa farmers utilized more of hired labour than family labour and less labour (weeding and pruning)

is required once the cocoa has been established, it could be acceptable. The farmers pay wages that are more than the value of their marginal production and hence would be inefficient as a result of allocative inefficiency (Idiong, 2006).

The negative coefficient of age of farm (Z_6) implies that efficiency decreases as the farms get older. Amos, (2007) had a similar result. Lastly the coefficient of farm size (Z_7) was positive and significant at the one percent level. This implies that technical efficiency increases with the size of farm. This result agrees with those of Giroh et al., (2008) and Amos (2007). Large farm sizes if properly managed should have higher efficiency. However, there is a threshold where returns to scale decreases with increase in farm size.

Elasticity of production and returns to scale

Typical of the power function (Cobb-Douglas), the estimated coefficients for the specified function can be explained as the elasticities of the explanatory variables.

The values of the variables indicates that a 10 percent increase in the size of farm, fertilizer, fungicide and labour will lead to a 1.5, 0.9, 4.2 and 0.9 percent increase respectively in

Table 7: Test of hypotheses that cocoa farmers in Cross River State are fully technically efficient ($Y= 0$)

Efficiency	Likelihood function (λ)	Log likelihood ratio (LR)	Critical X^2 0.05	Conclusion
Technical	-88.08	10.712	10.371	Reject

Source: Derived from Table 4. Critical X^2 was obtained from Kodde and Palm (1986).

Table 8: Frequency distribution of technical efficiency estimates

Efficiency level	Frequency	Percentage
0.20 – 0.30	2	1
0.31 – 0.40	3	1.5
0.41 – 0.50	20	10
0.51 – 0.60	17	8.5
0.61 – 0.70	38	19
0.71 - 0.80	91	45.5
0.81 – 0.90	26	13
>90	3	1.5
Total	200	100
Mean	0.69	
Minimum	0.20	
Maximum	0.93	

Source: Derived from output of the computer programme, Frontier 4.1 by Coelli (1994)

output of cocoa (Table 6). The value of the returns to scale (RTS) was 1.79 indicating that the farmers were operating in the region of decreasing but positive returns to scale (stage II of the production function). Therefore, increasing the units of inputs will not be the best option to the farmers as it will add less to total cocoa output (Table 7).

Technical efficiency estimates of the sampled cocoa farmers in Cross River Sate

The predicted farm specific technical efficiencies (TE) ranged between 0.20 and 0.93 (Table 8). The mean efficiency of the cocoa farmers was 0.69. The 69% mean efficiency indicates that in the short run, there is a scope of increasing cocoa production by about 31% by adopting the technologies and techniques practiced by the best cocoa farmers in the study area (Table 8).

The efficiency distribution also indicates that majority of the cocoa farmers (79%) were having efficiency of between 61% and 90% while a few of them (21%) were less than 60% efficient in their production process. The high levels of efficiency may be due to the long years of farming experience of the farmers.

CONCLUSION

This study reveals that, cocoa farmers in Cross River State are not fully technically efficient in their resource use. The policy variables that were identified as having significant effects (positive and negative) on the efficiency levels of the cocoa farmers are farmers age, family size, farm sizes, educational level, and age of the farm. It is believed that farmers' technical efficiency in resource use could increase since cocoa farmers in the area were not fully technically efficient; hence, there is room for improvement in the level of this efficiency if the important policy variables are addressed. Majority of farmers were in the productive age bracket and this was directly related to technical efficiency of cocoa farmers in the area. It is therefore important that a policy that would make cocoa farming attractive to persons within

this age bracket. The 69% mean efficiency indicates that in the short run, there is a possibility of increasing cocoa production by about 31% by adopting the technologies and techniques practiced by the best cocoa farmers in the study area.

REFERENCES

1- Adeniran, K. O (2007). Perspective on the role of Agriculture in meeting the Millennium Development Goal. World Bank Report.

2- Amos, T. T. (2007). An Analysis of Productivity and Technical Efficiency of Samallholder Cocoa Farmers in Nigeria. Journal of Social Science. 15 (2): 127-133.

3- Coelli, T. J. (1994). A Guide to Frontier 4.1: A Computer Programme for Stochastic Frontier Production. Department of Economics. University of New England, Amirdale.

4- Effiong, E. O. (2005). Efficiency of Production in Selected Livestock Enterprises in Akwa Ibom State, Nigeria. Unpublished Ph.D Dissertation, Department of Agricultural Economics, Michael Okpara University of Agriculture, Umudike, Nigeria.

5- Giroh, D. Y., Valla, W., Mohammed, A., & Peter, O. (2008). Analysis of the Technical Inefficiency of Gum Arabic Based Cropping Patterns among Farmers in the Gum Arabic Belt of Nigeria. Journal of Agriculture and Social Science. 4:125-128.

6- Idiong, I. C. (2006). Evaluation of Technical Allocative and Economic Efficiencies in Rice Production Systems in Cross River State, Nigeria. Unpublished Ph.D Thesis, Department of Agricultural Economics. Michael Okpara University of Agriculture, Umudike, Nigeria.

7- Kalirajan, K. (1981). The Economic Efficiency of Farmers Growing High Yielding, Irrigated Rice in India. American Journal of Agricultural Economics. 63(3): 566-569.

8- Kodde, F.C., & Palm, D.C. (1986). Wald Criteria for Jointly Testing Equality and Inequality Restrictions. Econometrical. 54: 1243- 1248.

9- Nkang, N. M., E. A. Ajah, S. O. Abang and E. O. Edet (2009). Investment in Cocoa Production in Nigeria: A Cost and Return Analysis of Three Cocoa Production Management Systems in Cross River State Cocoa Belt. African Journal of Food Agricultural Nutrition and Development. 2(2): 35-40.

10- Nwaru, J. C. (2004). Rural Credit Market and Arable Crop Production in Imo State of Nigeria. Unpublished Ph. D Dissertation, Michael Okpara University of Agriculture, Umudike, Nigeria.

11- Oluyole, K. A. and J. M. Usman (2006). "Assessment of Economic Activities of Cocoa Licensed Buying Agents (LBAs) in Odeda Local Government Area of Ogun State Nigeria". Akoka Journal of Technology and Science Education. 3(1): 130-140.

12- Oluyole. K. A. and R. A. Sanusi (2009). Socio-economic Variables and Cocoa Production in Cross River State Nigeria. Hum Ecol. 25 (1): 5-8.

13- Tadesse, B. and S. Krishnamurthy (1997). Technical Efficiency in Paddy Farms of Tamil Nadu: An Analysis Based on Farm Size and Ecological zone. Agricultural Economics. 16: 185 – 192.

14- Upton, M. (1972). Farm Management in Nigeria. Occasional Report Department of Agricultural Economics, University of Ibadan Nigeria.

5

Comparative Cost Structure and Yield Performance Analysis of Upland and Mangrove Fish Farms in Southwest, Nigeria

Mafimisebi Taiwo Ejiola[1] and Okunmadewa Foluso Yinka[2]

The bias against mangrove areas in siting fish farms prompted a comparison of the cost structure and yield performance in upland and mangrove locations. Tools utilized included descriptive statistics, budgetary and cash flow analyses and profitability ratios. Empirical results revealed that substantial revenue could be realized from both farms. While the upland farms yielded average gross revenue per hectare per year of $9,183.53, the mangrove farms made $8,135.93 revealing a slight difference. Results of combined cash flow and sensitivity analysis buttressed that of budgetary analysis. NPVs were $10,888.11 and $10,375.84, B/Cs were 1.28 and 1.29 and IRR were 48.55% and 48.51% for the upland and mangrove farms, respectively. Profitability ratios were also comparable but slightly higher in the upland farms. The conclusion is that there was little or no difference in yield performance. However, the high risk of investment loss in years of excessive flood should prompt investors in mangrove farms to compulsorily insure their farms.

Keywords:
Fish farming, Cost structure, Gross revenue, Investment yield, Upland and mangrove areas, Nigeria.

[1] *Department of Agricultural Economics and Extension, The Federal University of Technology, Akure, Nigeria.*
[2] *World Bank Country office, Asokoro, Abuja, Nigeria.*
 * *Corresponding author's email: temafimisebi@futa.edu.ng*

INTRODUCTION

The widening demand-supply gap for fish in Nigeria attests to the fact that there is the need to explore all avenues to increase and sustain fish supply. The factors implicated in the demand-supply deficit situation include water pollution from perpetual oil spillages which results in dwindling catches from capture fisheries, constant upward reviews of the prices of petroleum products which depress profit from capture fisheries, and over-fishing which involves large quantities of by-catch sold along with target species (Mafimisebi, 1995, FAO 2000 and Delgado, *et al.,* 2003). A right step towards arresting the demand-supply deficit for fish is aquaculture, which involves raising fish under controlled environment where their feeding, growth, reproduction and health can be closely monitored. Such farm-raised fish is already accounting for a considerable and rising proportion of total fish consumed in Nigeria and other developing countries (Delgado, *et al.,* 2003, Mafimisebi, 2007). The rapidly growing field of aquaculture has been recognized as a possible saviour of the over-burdened wild fisheries sector and an important new source of food fish for the poor (FAO 1995, Williams, 1996).

Most parts of the maritime (coastal) region of Nigeria (about 800km coastline, FDF, 1979) are suitable for aquaculture. The coastal area is in two parts; the upland communities (which make up about 25.0% of the total land area) and the mangrove swamp areas, which are perennially inundated by flood or flooded for most parts of the year. The upland communities are characterized by fresh water while the mangrove areas have brackish water. Aquaculture first started in the upland parts of the coastal areas of Southwest, Nigeria. Today, a good number of private commercial fish farms are found there. However, owing to the relative scarcity and high cost of acquiring land in the upland areas, many prospective fish farmers have not been able to commence the business of fish farming. In comparison, only a few private commercial fish farms are found in the relatively land-surplus mangrove parts. The particularly difficult terrain of the mangrove areas especially with respect to its physiographic nature, water quality, distance from input source, problem of fish pond construction and the possibility of flooding, were cited as reasons why prospective investors shy away from locating their fish farms in the mangrove areas (Mafimisebi, 1995).

The result of this is a vast expanse of mangrove land lying unused while there is serious competition for land in the upland parts. For example, Ondo State, one of the coastal states in Southwest, Nigeria, has an estimated 850 and 2450 hectares of exploitable fresh and brackish water fishing grounds. While more than 80.0% of the fresh water fishing grounds is being exploited with about 10.0% of this under aquaculture, less than 4.0% of the brackish water grounds is being exploited with only about 1.0% under fish farming (Ondo State Agricultural Development Project, 1996). In Nigeria as a whole, the same situation holds. Fish farming is the least exploited fishery sub-sector with the vast brackish water fishing grounds almost unexploited. Less than 1.0% of the fresh water grounds and about 0.05% of the brackish water grounds are under aquaculture to produce a current average yield of 20,500 tonnes of fish per annum. This represents only 3.12% of the estimated fish culture potential of 656,815 tonnes per annum. When the current output is compared with potential yield, one will immediately appreciate the need for increased effort at bringing most, if not all land suitable for aquaculture under cultivation (Ajayi and Talabi, 1984, Tobor, 1990, Falusi, 2003). In fact, apart from increasing the land area under aquaculture, astute management system targeted at doubling the present national aquaculture production rate of 1.5tonnes/ha/yr should be employed. If this can be achieved, total potential yield will increase to 1,831,000 tonnes per year, which will exceed the projected fish demand of between 1,562,670 tonnes and 1,609,920 tonnes per annum by the year 2010 and beyond (Tobor, 1990, Dada, 1996, FOS, 2005).

The observation that investors are biased against the mangrove areas of Southwest, Nigeria in siting of their fish farms, was the motivation to compare the yield performance and profitability in the two fish farm locations of upland and mangrove. This is necessary because commercial fish farmers have two major objectives; the provision of fish for human consumption and employment opportunities, which can only be realized when maximum income and profitability are achieved in farmed fish production (Fagbenro, 1987). The specific objectives of the study are to (1) describe the operational characteristics of the fish farms (2) compute and compare indices of yield performance in the two sets of farms and (3) identify the constraints encountered by farmers in the two locations.

MATERIALS AND METHODS
Study Area and Sampling Technique

The study was carried out in Southwest, Nigeria. Multi-stage sampling technique was used in collecting the data analyzed in this study. Two out of the six states that make up Southwest, Nigeria; Ondo and Ogun States, were purposively selected on the basis of having the highest aquaculture production figures. Four (4) Local Government Areas (LGAs), two from each state, were also purposively selected for having both upland and mangrove communities with exploitable fishing grounds. The two LGAs selected from Ogun State were Ijebu Waterside and Ijebu East while Ilaje and Ese-Odo LGAs were selected from Ondo State. From a list of registered commercial fish farms got from the Agricultural Development Programme (ADP) offices in the two states, thirty (30) and fifteen (15) commercial fish farms from the upland and mangrove areas were systematically selected.

Data and Data Collection

A well-structured questionnaire was used to obtain information on characteristics and management of fish farms, fish species cultured and items of costs and returns on the production process between years 2002 and 2006. Farm records of the farms surveyed were also made available to the data collectors. The fixed costs incurred in production were calculated as annual cost or rental values of fixed items. The depreciated cost (obtained through the straight-line method) represents annual lost in value of the facilities and equipment arising from wear and tear. The expected useful life (in years) of fixed items are indicated as follows: fish pond (20), boat/canoe (10), net (5), wheel-barrow (5), bowl (5), refrigerator/pumping machine (10), weighing scale (10), outboard engine (10), farm building (25) and hatchery (10). Copies of a set of questionnaires were administered to the owners or farm managers of the farms surveyed. The questionnaire was earlier pre-tested on fish farmers in the riverine areas of Irele and Ado-Odo/ Ota LGAs of Ondo and Ogun States respectively. In all, forty-five (45) fish farmers provided the data analyzed in this study.

Analytical Techniques

The data collected were analyzed using descriptive statistics which included frequency counts, percentages and tables. The budgetary model was used to determine the level of profit generated. The budgetary analysis was first carried out for all the five years (2002-2006) pooled together and then on a year-by-year basis. From the results of the yearly budgetary analysis, certain ratios of profitability and efficiency were obtained. These are:

i) Operating Ratio (OR) TVC/ GR   (1)
ii) Returns on Sales (ROS) NP/GR   (2)
iii) Returns on Assets (ROA) GM/ TCA (3)

Where TVC = Total Variable Cost, GR= Gross Revenue, NP = Net Profit, GM = Gross Margin and TCA = Total Cost of Assets.

The combined cash flow and sensitivity analysis was done to ascertain the extent of profitability of the aquaculture business and the factor(s) to which profitability is responsive. The profitability indicators used to measure the extent of returns from aquaculture are:

i) Benefit-Cost Ratio (B/C): This is the ratio of discounted costs to discounted revenue. A B/C of greater than unity is desirable for a business to qualify as a good one. Mathematically, B/C is stated as:

$$B/C = \frac{\sum_{t=1}^{n} \frac{Bt}{(1+r)^n}}{\sum_{t=1}^{n} \frac{Ct}{(1+r)^n}} \qquad (4)$$

where

Bt = benefit in each project year
Ct = cost in each project year
n = number of years
r = interest or discount rate

ii) Net Present Value (NPV): This is the value today of a surplus that a project makes over and above what it would make by investing at its marginal rate. Alternatively, it is defined as the value today of all streams of income which a project is to make in future. For a good business, NPV must be positive at the chosen discount factor. Mathematically, NPV is given as:

$$NPV = \sum_{t=1}^{n} \frac{Bt - Ct}{(1+r)^n} \qquad (5)$$

Where B_t, C_t, n and r are as earlier defined.

iii) Internal Rate of Return (IRR): It is the rate of return that is being expected on capital tied down after allowing for recoupment of the initial capital. The IRR is the rate of interest which equates the NPV of the projected series of cash flow payments to zero. It is also called the yield of an investment. Mathematically, it is given as:

$$IRR = \sum_{t=1}^{n} \frac{Bt - Ct}{(1+r)^n} = 0 \qquad (6)$$

Practically, the IRR is usually obtained through a series of manipulations where two discount factors give rise to two NPVs. The NPV must be positive at the lower discount factor and negative at the higher discount factor indicating that the project can earn higher than the lower discount factor and lower than the higher discount factor. In this trial and error method, according to Adegeye and Dittoh (1985), the IRR is given as:

$$IRR = \begin{array}{c} \text{lower discount} \\ \text{rate} \end{array} + \begin{array}{c} \text{difference between the two} \\ \text{discount rates} \end{array}$$

$$\left(\frac{NPV \, at \, lower \, discount \, rate}{absolute \, difference \, between \, the \, two \, NPVs} \right) \qquad (7)$$

RESULTS AND DISCUSSION
Operational and Farm Characteristics
The total farm size (area covered by fish ponds) of the 30 upland farms was 332,558m^2 while the total number of pond units was 168. Therefore, the average size of an upland fishpond was 1978m^2. The total size of the 15 mangrove farms was 1,226,575m^2 and the number of fishpond units was 154 giving an average size of 7,965m^2. Thus, fishponds have bigger sizes in the mangrove areas. The two possible reasons for this finding are that land is relatively cheaper in the mangrove areas and also that most mangrove farmers use the polyculture method while majority of upland farmers, use the monoculture method. About 96.0% and 68.2% of upland and mangrove farmers respectively, were engaged in purely table fish production while the balance in each case combined table fish with fingerlings/post-fingerlings (jumpers) production. The higher proportion of mangrove farmers combining both table fish and fish seeds production compared with the upland farms is probably owing to the presence of larger water bodies from which seeds of spawning fish can be harvested and further reared to jumpers before being used to stock ponds. This is a saving on cost of inputs but has a negative effect on capture fisheries since the fingerlings are the ones expected to grow into table fish in the natural water bodies (Touminen and Esmark, 2003).

Majority (60.0%) of the farmers procures their fish seeds from the wild (Table 1). According to the farmers interviewed, the natural source of fish seeds is cheaper and more readily available. The farmers in the upland parts contract out fish seeds procurement to people to whom they give part payment to facilitate timely delivery. Alternatively, fish farms sometimes assign that duty to some of their workers if they have enough workforce. In comparison, some workers of mangrove farms have fish seeds procurement as their major responsibility. They harvest fish seeds of various fish during the spawning seasons. Such harvested seed stocks are furthered reared in a special pond. During this period, stunted, deformed and unhealthy seed stocks are removed and the remaining used for stocking fish ponds. A higher proportion of mangrove farmers got their fish seeds from the wild. Procuring fish seeds from the wild by most mangrove farmers and some upland farmers, is an economic response to the problem of acute shortage of high quality fingerlings from government or private hatchery which is capable of crippling production. They claimed that where fish seeds produced in modern hatcheries are available, the cost is prohibitive more so because of the transportation cost incurred. Thus, despite the fact that farmers using fish seeds from the wild are aware of the negative impact of their action on yield from capture fisheries, they asserted that they will continue to use this source until there is an alternative arrangement that is acceptable to them. There had also been frequent bloody clashes between fish seeds harvesters and capture fisheries fisher folks in the study area. This means that an urgent alternative has to be found to procuring fish seeds from the wild if the problem of threatened stocks of wild fish reported by FAO (1995, 1998 and 2000) is not to be further compounded in Nigeria.

From Table 1, it is obvious that about 88.9% of farmers depend on the wild for their fish seeds. Only about 6.7% and 4.4% of farmers procure their fish seeds from own modern hatchery and government-owned hatchery. There is a need to re-orientate farmers away from using fish seeds from the wild to stock their fishponds. Most of the farmers which depend on the wild indicated that they would have preferred seed stocks from specialized private or public hatcheries because of their high quality if the prices can be reduced and if some hatcheries can be sited close to them.

Fish Species Cultured

The species commonly reared in the upland farms were Tilapia, Alestes, Heterotis and Catfish and these fish species were raised by over 70.0% of the farmers. The other fish species which are reared but not as frequently as the above species are Mudfish, Heterobranchus, Ophiocephalus, Aeroplane fish and Mormyrus. These species were reared by less than 25.0% of the sample farms. The reasons given for preference for the most popularly cultured species include (1) ease of procurement and high rates of survival of the seeds (2) easy culturing (3) fast growth and reproductive rates when supplementary feeding is practised (4) high yield and (5) high demand and price in the study area. In the mangrove farms, however, the commonest fish species reared in order of frequency are Alestes (78.13%), Tilapia (60.75%), Gymnarchus (56.25%) and Heterotis (46.88%). Other fish species were Catfish, Aeroplane fish and Ophiocephalus. Only about 30.0% of the sample mangrove farms had these fish species reared in them.

Cost Component Analysis for Fish Farms

In both farm situations, the single most expensive item of variable cost was fish feeds. It accounted for 51.43% of variable cost in upland farms and 57.79% in the mangrove farms. This difference is probably owing to the astuteness of the upland farmers who formed small groups of 5-10 and pooled their money together to buy fish feeds directly from feed milling companies whereas most mangrove farmers bought their feeds from the dealers in the headquarters of their LGAs. Travelling long distances to the headquarters of their LGAs to buy small quantities of feeds, each time they run out of stock, leads to increased transportation cost. Added together, seed stocks and fish feeds accounted for 68.73% and 69.67% of variable cost in the upland and mangrove farms respectively. This high proportion

Table 1: Distribution of Farmers by Source of Fish Seeds

Source of Fingerlings	Frequency	Percentage
From wild	27	60.00
From own modern hatchery	03	6.67
From other farmers rearing	13	28.89
wild fish seeds	02	4.44
From government hatchery	45	100.00
Total		

Source: Survey data, 2007

Table 2: Fixed and Variable Cost for One-Hectare Upland and Mangrove Fish Farms

Fixed Items	Upland Farms		Mangrove Farms	
	($)%	Composition	($)%	Composition
Land	335.54	4.10	120.36	1.10
Pond construction	872.41	10.67	875.57	8.08
Farm buildings	374.58	4.58	202.85	1.86
Vehicles + boats	979.64	11.98	1,761.66	16.18
Nets	183.54	2.25	624.54	5.74
Boreholes + Water pumps	1,814.98	22.53	1,103.20	10.13
Wheel Barrow + Basins	96.64	1.18	116.65	1.07
Generators/Deep freezers/ Weighing Scale	35.98	0.04	344.37	3.16
Local hatchery + Fencing materials	35.10	0.43	483.12	4.44
Labour (permanent)	3,419.83	41.83	5,252.95	48.24
Sub-total	**8,175.23**	**100.00**	**10,890.00**	**100.00**
Variable Items				
Fingerlings & Jumpers	5,721.47	17.30	3,549.06	11.88
Fish feeds	17,012.70	51.43	17,264.76	57.79
Fertilizer + Other Chemicals	8,740.02	2.64	1,614.42	5.40
Transportation + Fuelling	4,352.90	13.16	3,797.43	12.71
Repairs + Maintenance	3,034.72	9.18	2,479.21	8.30
Casual labour	2,081.96	6.29	1,171.51	3.92
Sub-total	**33,078.49**	**100.00**	**29,876.39**	**100.00**

Source: Survey data, 2007.
Note: For the period covered by the data used for this study, the average exchange rate was N127= $1 while the value fluctuated between N125 and N129.

accounted for by seed stocks and fish feeds as items of variable expenses is in accordance with the findings by Zadek (1984), Inoni, (1992) and Mafimisebi, (2003) that cost of feeds and seed stocks accounted for more than 50.0% of total production cost in aquaculture (Table 2).

The depreciated average fixed cost, per hectare of upland fish farm was $8,175.23 in the five years covered by the study. The corresponding value for the mangrove farm was $10,890.00. The depreciated cost of pond construction and vehicle/boats carried 10.67% and 11.98% respectively in the upland farms while the same items accounted for 8.08% and 16.18% respectively in the mangrove farms. The cost of pond construction was higher in the mangrove farms because the mangrove species have had to be cleared first before pond construction proper begins. Not only that, the problem of pond edge stabilization gulps a lot of money compared with upland ponds. This is more so because fish pond construction in the mangrove areas involves putting special structures in place to prevent escape of fish into the wild during slight or excessive flood. Table 2 also shows that land is cheaper in the mangrove areas. However, the cheap cost of land as an item of fixed cost is eroded by the heavy expense on pond construction in the mangrove parts.

For the two farms, labour was the single most expensive item of fixed cost. While labour accounted for 41.83% of fixed cost in the upland farms, the value for the mangrove farms was 48.24%. Also revealed in Table 2 is the fact that seed stocks constituted about 17.00% of variable cost in the upland farms while the corresponding value in the mangrove farms was 12.00%. This is attributable to the fact that more upland farmers than mangrove farmers procured seed stocks from sophisticated private and public hatcheries. The seed stocks bought from such hatcheries were more expensive than the ones bought from local hatcheries.

Gross Revenue

Gross Revenue (GR) is the amount realized from sale of table fish, fingerlings and jumpers. However, because revenue from sales of fingerlings and jumpers is negligible, only revenue from table fish production is considered in this study. The information on GR from the various fish species cultured is provided in tables 3 and 4.

While the upland farms made net revenue of $87,171.39 per hectare for the period studied, the mangrove farms got $81,446.06 per hectare for the same period. Therefore the average net profit per hectare per year was $9,183.53 and $8,135.93 in the upland and mangrove farms respectively. Table 3 shows that in the upland parts, Heterotis contributed the highest proportion of GR followed by Gymnarchus and Alestes while Ophiocephalus, Heterobranchus and Mormyrus together amounted to just about one-fifth of GR. It can thus be concluded that Heterotis, Gymnarchus and Alestes were the major commercial species cultured in the upland areas. Table 4 showed that Alestes accounted for the highest proportion of GR followed by Heterotis, Gymnarchus and Tilapia, in that order in the mangrove farms. Catfish, Ophiocephalus and Aeroplane fish contributed less than one-fifth of GR. The major commercial species in the mangrove parts were Alestes, Heterotis, Gymnarchus and Tilapia

For both farms, there was positive net revenue indicating that aquaculture is operating at a profit in the two locations. This finding of positive net revenue in the mangrove farms contradicts earlier findings by Inoni (1993), Falusi (2005) and Zadek (1984) that mangrove farms in Delta and Ogun States, Nigeria and Port Said, Egypt respectively, sustained losses during an operational period of between 1987 and 2003.

Profitability and Efficiency Ratios

The year-by-year results of the budgetary analysis are shown in Table 5. From the values given in the table, profitability ratios that enabled us to arrive at a conclusion as to the efficiency of operation of the two fish farms, were calculated.

The data presented in Table 6 showed the profitability ratios by year of the two sets of farms. A decrease in OR over time is an indication of a good and efficient business. A decline in OR in the study indicates either increasing TR

Table 3: Gross Revenue on a One-Hectare Upland Farm

Fixed Items	Gross Revenue per year ($)					
	2002	**2003**	**2004**	**2005**	**2006**	**Total 2002-2006**
Heterotis	4,336.18	4,711.20	4,701.75	5,475.05	6,544.36	25,768.54
Gymnarchus	3,869.69	3,710.66	4,152.84	4,475.54	5,565.61	21,774.34
Alestes	2,543.66	1,687.22	3,150.56	3,799.76	4,302.56	15,483.76
Tilapia	1,146.18	1,100.48	1,425.83	1,297.55	1,333.61	6,309.38
Catfish	905.79	1,020.44	1,267.51	1,529.09	1,705.11	6,427.94
Heterobranchus	735.33	821.90	1,170.59	1,261.75	1,604.48	5,594.05
Mudfish	248.05	185.26	321.26	401.57	354.33	1,510.47
Ophicephalus	659.29	584.21	818.12	924.16	1,314.05	4,299.83
Mormyrus	0.44	0.32	0.57	0.69	1.06	3.07
Total	**14,444.61**	**13,821.70**	**17,014.75**	**19,165.17**	**22,725.17**	**87,171.39**

Source: Survey data, 2007.

Table 4: Gross Revenue on a One-Hectare Mangrove Farm

Fixed Items	Gross Revenue per year ($)					
	2002	**2003**	**2004**	**2005**	**2006**	**Total 2002-2006**
Heterotis	3,032.48	2,869.13	3,618.94	3,834.44	4,466.87	17,821.87
Gymnarchus	2818.32	2,583.13	3,276.63	3,784.87	3,961.48	16,424.43
Alestes	3,830.15	3,520.86	4,125.83	4,821.94	5,624.02	21,922.80
Tilapia	1,926.61	1,637.28	2,191.20	2,792.60	3,065.33	11,613.02
Catfish	1,529.54	1,338.63	1,710.50	1,993.95	2,487.39	9,060.16
Ophiocephalus	733.53	618.66	867.52	1,009.74	1,321.36	4,550.81
Aeroplane fish	6.90	6.02	9.63	14.01	16.56	53.11
Total	**13,877.53**	**12,573.72**	**15,800.24**	**18,251.54**	**20,943.02**	**81,446.06**

Source: Survey data, 2007.

Table 5: Year by Year Budgetary Analysis of a One-Hectare Farm

	Total Variable Cost ($)		Gross Revenue ($)		Gross Margin ($)		Net Profit ($)	
Year	Upland	Mangrove	Upland	Mangrove	Upland	Mangrove	Upland	Mangrove
2002	6,456.50	5,795.91	14,365.83	13,877.53	7,988.07	8,081.17	7,608.72	6,931.37
2003	5,481.22	5,090.61	13,821.70	12,573.72	8,340.48	7,483.11	7,280.60	6,280.16
2004	5,244.86	4,612.34	17,014.75	15,800.24	11,769.89	11,187.90	8,962.55	7,891.71
2005	7,272.51	6,695.12	19,165.17	18,251.54	11,892.66	11,556.44	10,095.28	9,116.05
2006	8,623.40	7,682.41	22,725.16	20,943.02	14,101.76	13,260.62	11,970.52	10,460.36

Source: Survey data.

Table 6: Profitability and Efficiency Ratios for a One-Hectare Farm

	Operating ratios		Return on sales		Return on sales	
Year	Upland	Mangrove	Upland	Mangrove	Upland	Mangrove
2002	0.447	0.418	0.527	0.499	1.680	1.434
2003	0.397	0.405	0.527	0.499	1.754	1.327
2004	0.308	0.292	0.527	0.499	2.475	1.985
2005	0.379	0.367	0.527	0.499	2.501	2.050
2006	0.379	0.367	0.527	0.499	2.965	2.352
Average	**0.382**	**0.370**	**0.527**	**0.499**	**2.275**	**1.830**

Source: Survey data.

or decreasing TVC. For the upland farms, OR was 0.447 in 2002 which decreased to 0.397 and 0.308 in 2003 and 2004. The value took an upward turn to 0.379 which was maintained in 2006. The same pattern was observed in the mangrove farms. OR fell from 0.418 in 2002 to 0.405 and 0.292 in 2003 and 2004 respectively but picked up to 0.367 in 2005 which remained same in 2006. For the period studied, average OR was 0.382 in the upland farms and 0.370 in the mangrove farms. Judging by these ratios, the mangrove farms seemed to promise a better efficiency in future years as OR was falling faster than in the upland farms. The increase in OR in years 2005 and 2006 on both farms is clearly not a desirable situation. The farmers must do all that is possible to achieve a consistently decreasing OR. This can be achieved by a more efficient use of farm resources. For example, feeding fish beyond a stipulated market weight should be avoided as the rate of growth slows down compared with the quantity of feeds consumed. Also, farmers should explore avenues for wider market outlets so that mature fish can be promptly disposed off. This scenario will lead either to a decreasing TVC or an increase in TR which will depress OR.

An increasing return on sales over time indicates a stable, profitable and efficient business. Return on sales was constant in the period studied on both farms. It was 0.527 for the upland farms and 0.499 for the mangrove farms. Fish farmers in both farm locations need to take steps to ensure an increasing return on sales.

The indication that assets are being more increasingly utilized is increasing returns on assets. On the upland farms, there was an increasing trend of returns to assets from 2002 to 2005 but there was a fall in the value in 2006. On the mangrove farms, the trend in returns on assets was towards an increase except in years 2003 and 2006 in which the figures fell below the year preceding them.

Combined Cash Flow and Sensitivity Analysis

Some assumptions were necessary in carrying out this analysis. These assumptions are as follows:

(1) The average bank lending rate to agriculture in the thirteen (13) years covered by the analysis is 25.0%.

(2) A risk-discounted factor of 5.0% is added to the bank lending rate meaning that a discount factor (DF) of 30% is used.

(3) There is a 20.0% and 10.0% projected annual increase in variable cost and unit price of fish between 2006 and 2014. This is in accordance with the farm management maxim which says it is better to be optimistic about cost rise and pessimistic about revenue increase in the estimation of future profitability of a business (Adesimi, 1985).

The result of the combined cash flow and sensitivity analysis for the upland and mangrove farms are shown in Tables 7 and 8 respectively. The results indicate that aquaculture is profitable at both locations at the assumed bank lending rate in spite of prices of key production inputs rising faster than output price. For the upland farms, the NPV stood at $10,887.24, the B/C was 1.28 and IRR was 48.55%. The corresponding values for the mangrove farms were $10,375.84, 1.29 and 48.51%. Thus, the results are comparable and do not show any considerable difference in yield performance between the two types of farms. While at the assumed bank lending rate, the upland farms would return $0.19 for every $0.79 invested, the farms located in the mangrove areas will also return approximately $0.19.

Constraints to Upland and Mangrove Fish Farming

Fish farmers in the two farm locations were asked to rank the constraints identified in their business. The problems encountered by the mangrove farmers in rank order were (1) financial constraints; (2) high and rising cost of feeds; (3) flooding which leads to total loss of investment whenever it happens as fish escape into the wild. Numerous studies have named potentially negative effects of escaped farmed fish on wild populations (Naylor et al., 2000); (4) silting up of ponds which result in massive death of cultured fish; (5) pests which include snakes, water-dogs and piscivorous birds; (6) attack by capture fishermen during sourcing of fish seeds from wild; (7) water pollution and (8) inadequate access to extension services. Only about 30% of mangrove farmers had had a contact with extension agents since commencement of business.

The problems commonly encountered by farmers in the upland areas were (1) financial constraints occasioned by high running costs; (2) drying up of ponds owing to seepage of water through dykes; (3) massive loss of fish owing to polluted or high-temperature water; (4) scarcity of high quality seed stocks and (5) problems of

Table 7: Cash Flow and Sensitivity Analysis for a One-Hectare Upland Farm (2002-2014)

Year	Cost ($)	Revenue ($)	Incremental Benefit ($)	DF 30%	NPV 30% ($)	DF 50%	NPV 50% ($)	Discounted Cost ($)	Discounted Revenue ($)
2001	21,774.18	-	-21,774.18	0.769	-16,744.35	0.667	-14,523.38	16,744.35	-
2002	6,456.50	14,444.57	7,988.07	0.592	4,728.94	0.444	3,546.71	3,822.25	8,551.19
2003	5,481.22	13,821.70	8,340.48	0.445	3,711.51	0.296	2,468.78	2,439.14	6,150.66
2004	5,244.86	17,014.74	11,769.89	0.350	4,119.46	0.198	2,330.44	1,835.70	5,955.16
2005	7,272.51	19,165.17	11,892.66	0.269	3,199.13	0.132	1,569.83	1,956.30	5,155.43
2006	8,623.40	22,725.17	14,101.76	0.207	2,919.06	0.088	1,240.96	1,785.04	4,704.11
2007	10,348.08	24,997.45	14,649.37	0.159	2,329.25	0.059	864.31	1,645.35	3,974.59
2008	12,417.70	27,497.19	15,079.49	0.123	1,854.78	0.039	588.10	1,527.38	3,382.15
2009	14,901.24	30,246.91	15,345.67	0.094	1,442.49	0.026	398.99	1,400.72	2,843.21
2010	17,881.49	33,271.60	15,390.12	0.073	1,123.48	0.017	261.63	1,305.51	2,428.83
2011	2,142.82	36,598.76	15,140.98	0.056	847.89	0.012	181.69	1,201.64	2,049.53
2012	2,576.11	40,258.64	14,509.30	0.043	623.90	0.008	116.07	1,107.22	1,731.12
2013	30,899.21	44,284.50	13,385.29	0.033	441.71	0.005	66.93	1,019.67	1,461.39
2014	37,079.47	48,712.95	11,633.90	0.025	290.85	0.003	34.90	926.98	1,217.84
					10,888.11		**-854.04**	**38,717.24**	**29,031.78**

Source: Field data and projected figures

Notes: (1) 2001 is the investment year (year zero), so there is no revenue

(2) Costs and Revenues for 2002-2006 are actual flows recorded by the fish farms

(3) Cost and revenues for 2007 – 2014 are projected figures

NPV at 30% = 10,888.11   IRR = 48.55%   B/C = 1.28

Table 8: Cash Flow and Sensitivity Analysis for a One-Hectare Mangrove Farm (2002-2014)

Year	Cost ($)	Revenue ($)	Incremental Benefit ($)	DF 30%	NPV 30% ($)	DF 50%	NPV 50% ($)	Discounted Cost ($)	Discounted Revenue ($)
2001	20,898.93	-	-20,898.93	0.769	-16,071.28	0.667	-9,215.18	16,071.28	-
2002	5,795.91	13,877.53	8,081.62	0.592	4,784.32	0.444	3,588.24	3,431.18	8,215.50
2003	5,090.61	12,573.72	7,483.11	0.445	3,329.98	0.296	2,215.00	2,265.32	5,595.30
2004	4,612.34	15,800.24	11,187.90	0.350	3,915.77	0.198	2,215.20	1,614.32	5,530.09
2005	6,695.12	18,251.54	11,556.44	0.269	3,108.68	0.132	1,525.45	1,800.98	4,909.67
2006	7,682.41	20,943.02	13,260.62	0.207	2,744.95	0.088	1,166.93	1,661.12	4,335.21
2007	9,218.73	23,037.33	13,818.44	0.159	2,197.13	0.059	815.29	1,465.80	3,662.93
2008	11,062.67	25,341.06	14,278.39	0.123	1,756.24	0.039	556.82	1,360.71	3,116.95
2009	13,275.20	27,875.16	14,599.96	0.094	1,372.40	0.026	379.60	1,247.87	2620.27
2010	15,930.24	30,662.68	14,732.44	0.073	1,075.47	0.017	250.45	1,162.91	2,238.39
2011	19,116.29	30,662.68	14,612.66	0.056	818.31	0.012	175.35	1,070.51	1,888.82
2012	22,939.55	33,728.95	14,162.30	0.043	608.98	0.008	113.30	986.40	1,595.38
2013	27,527.46	37,101.84	13,284.57	0.033	438.39	0.005	66.42	908.41	1,346.80
2014	33,032.95	40,812.13	11,860.28	0.025	296.51	0.003	35.58	825.82	1,122.33
					10,375.84		**-835.92**	**35,801.77**	**46,177.60**

Source: Field data and projected figures
Note: NPV at 30% =10,375.84   IRR = 48.51%   B/C = 1.29

theft which can lead to over-night harvesting of fish ponds with marketable fish if security is not beefed up around ponds. These are the problems that must have solutions proffered to them for the operation of these farmers to be enhanced.

CONCLUSION AND RECOMMENDATIONS

The study explored the operational characteristics of upland and mangrove farms, compared costs and profitability of investment in the two locations and determined the production variables to which profitability is more sensitive. The study also examined the constraints to fish farming in the two locations.

Empirical results show that mangrove farms are about four times bigger in size than the upland farms. Monoculture method was prevalent among mangrove farmers while the upland farmers mostly practised polyculture. The commercial species reared in the study area were Tilapia, Alestes, Heterotis and Gymnarchus while Ophiocephalus, Catfish, Heterobranchus and Mormyrus were the minor commercial species. The depreciated fixed cost in the upland farms was lower than that of the mangrove farms. However, the level of variable cost in the upland farms was greater than that of the mangrove farms. In both farms, the cost of labour was the single most expensive item of

variable cost. G.R per hectare was comparable in the two farm situations but slightly higher in the upland farms. Profitability ratios which indicate efficiency did not show any considerable difference in the yield of investment from the two farm locations. The result from the combined cash flow and sensitivity analysis shows that investment in aquaculture is profitable at both farm locations. All performance indicators show that profitability is not different between farms in the two locations to justify the avid preference for the upland locations in the siting of fish farms in the coastal areas of Southwest, Nigeria.

The magnitude of cost involved in establishing and managing a fish farm is clearly beyond that affordable by a peasant farmer. Investible funds in form of loans should be made available to prospective investors wishing to site their farms in the mangrove areas at affordable interest rate. The study has shown that fish farmers can repay loans advanced to them conveniently if given a moratorium of two years.

There is also the need to encourage investors in hatcheries to produce fish seeds for use by fish farmers especially in the mangrove areas where majority of the farmers depend on the wild for their fish seeds. Once the government has succeeded in attracting investors in hatcheries to the mangrove areas, a campaign against the use of fish seeds from the wild in the study area should be launched. As soon as the hatcheries can produce enough seed stocks to satisfy all identified fish farmers, the practice of procuring fish seeds from the wild should be banned. This is a matter of priority if yield from aquaculture is to be increased and natural fisheries resources conserved. Finally, since the level of capital investment for establishing fish farms is very high, the government can subsidize cost of fish feeds for new investors in the mangrove areas only in the first year of operation. This may serve to attract investors into the area so that the vast mangrove land can be put to productive use. There is also the need to step-up extension visits to fish farmers.

It is also recommended that farmers in the mangrove areas take policy with the Nigerian Agricultural Insurance Company so that they can be indemnified if there is loss of investment from fish escapes during periods of excessive flood.

All performance indicators show that profitability is not different between farms in the two locations to justify the bias against the mangrove areas in the siting of fish farms in the coastal areas of Southwest, Nigeria. Solving some of the identified problems of fish farmers in both locations is a step towards cheap and affordable animal protein production especially in the vast mangrove areas with its hydrographic characteristics.

ACKNOWLEDGEMENT

Thanks are due to Dr. I. Ajibefun of The Federal University of Technology, Akure, Nigeria, Professor Yoshi Matsuday and Ms. Ann Shriver, and two other anonymous reviewers who are members of International Institute of Fisheries Economics and Trade (IIFET) that reviewed developing countries' aquaculture economics papers for various awards and travel grants for the 2006 conference in Portsmouth, U.K. IIFET is gratefully acknowledged for a partial travel grant of $900 to enable the first author attend and give a work-in-progress version of this paper in IIFET 2006, Portsmouth. Various other members of the Aquaculture Economics Session in the said conference are thanked for raising useful comments that have been addressed in this revised version of the paper.

REFERENCES

1- Adegeye, A.J. & Dittoh, J.S. (1985). Essentials of Agricultural Economics. Impact Publishers Nig. Ltd., Ibadan. p 251.

2- Adesimi, A.A. (1985). Farm Management Analysis. Tintner Publishing Company, pp 105-206.

3- Ajayi, T.O. & Talabi, S.O. (1984). The Potentials and Strategies for Optimum Utilization of the Fisheries Resources of Nigeria. NIOMR Technical Paper, No. 18, P 24.

4- Dada, B.F. (1976). Present Status and Prospects for Aquaculture in Nigeria. NIOMR Technical Paper 52: 5-12.

5- Delgado, C.L., Wada, N., Rosegrant, M.W., Meijer, S. and M. Ahmed. (2003). Fish to 2020: Supply and Demand in Changing Global Markets. International Food Policy Research Institute, Washington D.C, World Fish Center, Penang, Malaysia P 223.

6- Fagbenro, O.A. (1987). A Review of the Biological and Economic Principles Underlying Commercial Fish Culture. Journal of West African Fisheries, 11(2): 171-177.

7- Falusi, O.A. (2003). Poverty Analysis of Fish Farmers in Abeokuta Metropolis. An Unpublished Project Report, Department of Agricultural Economics and Farm Management, University of Agriculture, Abeokuta, Nigeria. p 57.

8- FAO (1995). The State of World Fisheries and Aquaculture, Rome.

9- FAO (1998). The State of World Fisheries and Aquaculture, Rome.

10- FAO (2005). The State of World Fisheries and Aquaculture, Rome.

11- Federal Department of Fisheries (1979). Fisheries Statistics of Nigeria. pp 1-32.

12- FOS (2000) Projected Fish Demand and Supply in Nigeria (1999-2010). Federal Office of Statistics, Abuja, Nigeria.

13- Inoni, O.E.,(1992). Financial Analysis of Fish Farming in Delta State, Nigeria. Unpublished M.Sc Dissertation, University of Ibadan. p 88.

14- Mafimisebi, T.E. (1995). Profitability and Yield Performance of Selected Fish Ponds in Ilaje Ese-Odo Local Government Area of Ondo State, Nigeria. Unpublished M.Sc Dissertation, University of Ibadan. p. 83.

15- Mafimisebi, T.E. (2003). Yield Performance of Commercialized Upland Fish Farms in Ondo State of Nigeria. Nigerian Journal of Animal Production. 30 (2): 217-228.

16- Naylor, R.L., Goldburg, R.J., Primavera, J.H., Kautsky, N., Beveridge, M.C.M., Clay, J., Folke, C., Lubchenco, Mooney J. H., & Troell, M. (2000). Effects of Aquaculture on World Fish Supplies. Nature 405: 1017-1024.

17- Ondo State Agricultural Development Project (1996). Aquaculture Potentials of Ondo State of Nigeria. pp 1-10.

18- Tobor, J.G.(1990). The Fishing Industry in Nigeria: Status and Potentials for Self-Sufficiency in Fish Production. NIOMR Technical Report. No. 54: 1-23.

19- Touminen, T.R., & Esmark, M. (2003). Food for Thought: The Use of Marine Resources in Fish Feeds. WWF Report 02/03. Norway: World Wildlife Fund.

20- Williams, M.J. (1996). Transition in the Contribution of Living Aquatic Resources to Sustainable Food Security". In Perspectives in Asian Fisheries ed Sena S. De Silva. Makati City, The Philippines: Asian Fisheries Society.

21- Zadek, S.,(1984) Development de l'Aquaculture in Egypte. Reference a la Farme de Reswa (Port Said et Froposition d'une Politique National Equacole. Ph.D Thesis, Toulouse Institut National Polytechnique de Toulouse P. 151.

6

Effectiveness of Extension Services in Enhancing Outgrowers' Credit System: A Case of Smallholder Sugarcane Farmers in Kisumu County, Kenya

Abura Odilla Gilbert [1], Barchok Kipngeno Hillary[1] and Onyango Christopher Asher[2]

Keywords:
Public extension service, Private extension service, Outgrowers' Credit system, Effectiveness, Western Kenya

The purpose of this study was to investigate the role of extension services in enhancing effectiveness of outgrowers' credit system in Kisumu County, Kenya. The study specifically sought to determine whether public and private extension services play a significant role in enhancing effectiveness of out-growers' credit system among smallholder sugarcane farmers. A total of 110 small scale farmers were randomly selected for the study. A closed ended questionnaire was used to collect data from farmers. Both descriptive and inferential statistics were used for data analysis. The findings indicated that both public and private extension services were insignificant in enhancing effectiveness of outgrowers' credit system. Further, the findings indicated that there was no significant difference between public and private sector in provision of extension services. The findings suggest that for outgrowers' credit system to be effective in terms of creation of awareness about credit, accessibility, timely supply of credit, supervision of credit and provision of extension advice on credit utilization, both public and private extension services should be intensified and coordinated to avoid duplication. The results also suggest that sugarcane factory extension division should be strengthened just like in the coffee and tea sub-sectors.

[1] *Chuka University College, P. O. Box 109, Chuka, Kenya.*
[2] *Egerton University, P. O. Box 536, Egerton, Kenya.*
* *Corresponding author's email: gilbura@yahoo.com*

INTRODUCTION

The importance of agriculture to the African economies is stressed due to the fact that agriculture remains the principal occupation of the majority of people, constitutes the largest production sector, and produces an average of 32% of GDP, major sources of raw materials for industries and a significant purchaser of the countries manufacturers and services (Agbamu, 2005). In Kenya for example, the economy is heavily dependent upon agricultural sector and as the World Bank (2007) report indicates, the country's future will considerably depend on productivity of smallholder farms. Agriculture is by far the single largest economic sector in Kenya and accounts for about 30% of GDP, over 60% of the exports, 75% of the total labour force and provides 80% of industrial raw materials (Economic Survey, 2007; Kenya Sugar Research Foundation, 2007, Government of Kenya, 2005). Since independence, smallholder agriculture gained ground from mere provision of subsistence and minimal marketed surplus to account for over three quarters of agricultural production and 85% of agricultural employment (GoK, 2005, World Bank, 2007). Sugarcane farming is one such subsector that contributes to the national economy. According to Guda et al., (2001) smallholder farmers accounts for 89% of the total area under sugarcane farming in Kenya. This provides an investment opportunity. However, this is only possible if the problems affecting it are addressed. Some of the main problems include shortage of sugarcane due to lack of systemic and synchronised sugarcane development, poor crop husbandry practices, poor cane varieties and qualities in some factory zones, poor harvesting methods, poor management in some factories affecting factory efficiency and output and inadequate agricultural production credit among others (Guda et al., 2001). However the most notable problem is complaints due to delayed payment dues to the farmers after cane delivery. These problems have elicited diverse reactions from the farmers. The most severe reactions are cases of burning the cane crop by the farmers in their own farms, so as to turn to other lucrative enterprises like maize or bean seed production (Agribusiness Development Support Project Annual Report, 2001).

Agricultural extension is considered to be an important service in increasing agricultural productivity and attaining sustainable development (Kibet, et al., 2005). Its role is to help people identify and address their needs and problems. There is a general consensus that extension services if successfully applied, should result in outcomes which include observable changes in attitudes and adoption of new technologies, and improved quality of life based on indicators such as health, education and housing. It has been recognized that agricultural extension accelerates development in the presence of other factors such as markets, agricultural technology, availability of supplies, production incentives and transport (Kibet, et al., 2005). Koyenican (2008) equates help in extension to empowering all members of the farm households to ensure holistic development. This is because agricultural extension brings about changes, through education and communication in farmers attitude, knowledge and skills.

The performance of the public agricultural extension service in Kenya has been a very controversial subject (Gautam and Anderson, 1999). The system has been perceived as topdown, uniform (one-size-fits-all) and inflexible and considered a major contributor of the poor performing agricultural sector (Government of Kenya, 2005). Thus there has been a desire to reform extension in to a system that is cost effective, responsive to farmer's needs, broad based in service delivery, participatory, accountable and sustainable. As a result of ineptness in the public extension system, private agricultural extension system has emerged comprising of private companies, non-governmental organizations (NGO's), community based organizations (CBO's) and faith based organizations (Nambiro et al., 2005 and Rees et al., 2000).

Agricultural extension as a public sector institution has an obligation to serve the needs of all agricultural producers, either directly or indirectly (Anderson, 2007). This is because public sector extension is a public good. The Kenya government has tried a number of extension models and styles, including the progressive (farmer approach) model, integrated agricultural rural development approach, farm management, training and visit, farming systems approaches and farmer field schools. All these approaches have emerged with varying level of success for different groups. However, the effectiveness of extension services in enhancing effectiveness of outgrowers' credit system among sugarcane farmers in Kisumu County of Kenya has not been examined. Thus the present study was set with the premise that both outgrowers' credit and extension service are instruments for promoting agricultural development and that an efficient and effective extension service is important in enhancing effectiveness of outgrowers credit system. Credit to farmers is an important instrument in improving productivity. Indeed as Wangia (2001) noted, it is a prerequisite to the adoption of improved agricultural technologies for the smallholder farmers. Nevertheless, for credit system to help the smallholder farmers it should be tied to improved technologies, remunerative prices for the farmers' output and good extension network (Ogunsumi, 2004).This paper presents results on the role of extension services in enhancing effectiveness of outgrowers' credit system among smallholder cane farmers in Kenya.

Conceptual Framework

This study formulated a conceptual model that encompassed major variables and their possible patterns of influence on each other and eventually on effectiveness of outgrowers' credit system. The effect of the extension services namely public and private services are mediated by level of education, farm size and farmers period of residence. What this structural model indicates therefore is that the moderator variables influence adoption of sugarcane technologies disseminated by extension agents whether from public or private sector. In view of this model, the theory underpinning this study is that, adoption is complex and multifaceted process. While the main activity of extension centers on increasing production, this study concentrated on implementation of such activities. These are; creation of awareness about credit, accessibility to credit, timely supply of credit, supervision of credit and provision of extension advice on credit utilization (Table 1).

MATERIALS AND METHODES

Ex-post facto survey design was adopted for this study. Kisumu county in Kenya was purposely selected because of its uniqueness in that the county boasts of three major sugar factories.

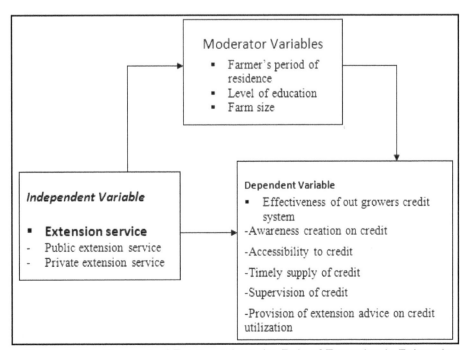

Figure 1: The Conceptual Framework on the Role of Extension in Enhancing Effectiveness of Outgrowers' Credit System

Table 1: Creation of Awareness on credit facility

Activity		SD	D	U	A	SA	Total
Creation of Awareness	f (n=23)	2.0	17.0	-	2.0	2.0	23.0
	%	8.7	73.9	-	8.7	8.7	100.0

Key (SD= Strongly Disagree, D= Disagree, U= Undecided, A= Agree, SA= Strongly Agree)

Table 2: Land Preparation

Activity		SD	D	U	A	SA	Total
Land Preparation	f (n=23)	2.0	14.0	-	6.0	1.0	23.0
	%	8.7	60.2	-	26.0	4.3	100.0

Key (SD=Strongly Disagree, D=Disagree, U=Undecided, A=Agree, SA= Strongly Agree)

These are; Miwani, Muhoroni and Chemelil which were established in the years 1923, 1966 and 1968 respectively. The County has favourable moderate climatic conditions, with temperatures averaging 27° C and receives bimodal rainfall ranging from (560 -1630) mm per annum. Kisumu County comprises of the main topographical land formations namely, the Nandi hills, the Nyando plateau and Kano plains which are sandwiched between two hills. The Kano plains comprise predominantly black cotton clay soils derived from igneous rocks. The County's altitude range from 1000-1860 M above sea level. The target population was the sugarcane farmers in Kisumu county. A total of 110 smallholder cane farmers were randomly selected for the study but only 108 farmers questionnaire were useful for analysis. A closed ended questionnaire was used to collect data by personal interviews. The information gathered was analysed using both descriptive and inferential statistics.

RESULTS AND DISCUSSIONS
Public extension Services

Tables 1, 2 and 3 below shows data on different extension activities done by public sector extension service for cane farmers. Table 1 show result on the effect of public extension services in relation to creation of awareness. Creation of awareness was one of the activities used to measure public extension services. To elicit information on creation of awareness, the farmers were asked to respond to statement designed to elicit negative responses on performance resulting to creation of awareness. A 5-point likert scale was constructed to record these responses. Table 1 show results on public extension services in

Table 3: Appropriate Input Use

Activity		SD	D	U	A	SA	Total
Appropriate Input Use	f (n=23)	2.0	7.0	6	6.0	2.0	23.0
	%	8.7	30.4	26.0	26.0	8.7	100.0

Key (SD=Strongly Disagree, D=Disagree, U=Undecided, A=Agree, SA=Strongly Agree)

Table 4: Creation of awareness on credit facility

Activity		SD	D	U	A	SA	Total
Creation of Awareness	f	16.0	51.0	-	9.0	4.0	80.0
	%	20.0	63.8	-	11.3	5.0	100.0

Key (SD=Strongly Disagree, D=Disagree, U=Undecided, A=Agree, SA=Strongly Agree)

Table 5: Land Preparation

Activity		SD	D	U	A	SA	Total
Land Preparation	f (n=80)	9.0	12.0	1.0	35.0	23.0	80.0
	%	11.25	15.0	1.25	43.8	28.8	100.0

Key (SD=Strongly Disagree, D=Disagree, U=Undecided, A=Agree, SA=Strongly Agree)

Table 6: Use of Appropriate Inputs

Activity		SD	D	U	A	SA	Total
Timely Use of Appropriate Inputs	f (n=80)	3.0	17.0	-	44.0	16.0	80.0
	%	3.75	21.3	-	55.0	20.3	100.0

Key (SD=Strongly Disagree, D=Disagree, U=Undecided, A=Agree, SA=Strongly Agree)

relation to creation of awareness.

The results in Table 1 indicated that majority of farmers (82.6%) receive information on credit facility from public extension providers.

Advice on land preparation was one of the activities used to measure public extension services. To elicit information on land preparation, the farmers were asked to respond to statement designed to elicit positive knowledge of performance resulting to land preparation. A 5-point likert scale was constructed to record these responses. Table 2 show results on public extension services in relation to land preparation.

The result in table 2 showed that land preparation as an activity has not been satisfactorily addressed by public service extension officers as reflected by the high percentage (68.9%) of farmers who disagreed with the positive statement. This suggests that perhaps the declining cane production in Kisumu County is due to poor and inadequate land preparation, culminating from inadequate machinery to prepare land for cane growing.

Advice on appropriate use of inputs was one of the activities used to measure public extension services. To elicit information on the use of appropriate input, the farmers were asked to respond to statement designed to elicit positive knowledge of performance resulting to use of appropriate input. A 5-point likert scale was constructed to record these responses. Table 3 show results on public extension services in relation to appropriate use of input.

The results in table 3 showed that 39.1% of the farmers have not received services on appropriate use input from public sector extension services. This suggests that farmers do not know whether the inputs they use are appropriate or not.

Private extension service

Private sector extension may play a predominant extension role for particular inputs, particular enterprises / commodities and for particular farmer's in particular geographical areas. This enables farmers to benefit from increased incomes and economic security. Tables 4, 5, and 6 shows data on different extension activities done by private sector extension service for cane farmers.

Table 4 show result on the effect of private extension services in relation to creation of awareness. Creation of awareness was one of the activities used to measure private extension services. To elicit information on creation of awareness, the farmers were asked to respond to statement designed to elicit negative knowledge of performance resulting to creation of awareness. A 5-point likert scale was constructed to record these responses. Table 4 show results on private extension services in relation to creation of awareness.

The results in table 4 showed that majority of farmers (83.75%) received information on credit facility.

Advice on land preparation was one of the activities used to measure private extension services. To elicit information on land preparation, the

Table 7: Comparison between role of public and private extension services in Kisumu County

		Public Extension Service (N=23)				Private Extension Service (N=80)			
		LP	UAI	CA	Average	LP	UAI	CA	Average
SA	f	1.0	2.0	2.0	2.0	23.0	16.0	4.0	13.2
	%	4.3	8.7	8.7	8.7	28.8	20.3	5.0	16.5
A	f	6.0	6.0	2.0	5.0	35.0	44.0	9.0	28.2
	%	26.0	26.0	8.7	21.5	43.8	55.0	11.3	35.3
U	f	-	6.0	-	2.0	1.0	-	-	00.8
	%	-	26.0	-	8.7	1.25	-	-	01.0
D	f	14.0	7.0	17.0	12.5	12.0	17.0	51.0	27.4
	%	60.2	30.4	73.9	53.8	15.0	21.3	63.8	34.3
SD	f	2.0	2.0	2.0	1.8	9.0	3.0	16.0	10.4
	%	8.7	8.7	8.7	7.74	11.25	3.75	20.0	13.0
Total	f	23.0	23.0	23.0	23.0	80.0	80.0	80.0	80.0

Key (SD=Strongly Disagree, D=Disagree, U=Undecided, A= Agree, SA=Strongly Agree)
LP- Land Preparation
UAI- Use of Appropriate Inputs
CA- Creation of Awareness

farmers were asked to respond to statement designed to elicit positive knowledge of performance resulting to land preparation. A 5-point likert scale was constructed to record these responses. Table 5 show results on public extension services in relation to land preparation.

The results in 5 showed that majority of the farmers (72.5%) received extension services on land preparation from private sector. Perhaps this is due to the diverse nature of private sector extension where land preparation machines are readily provided to try and promote return on investment as well as enabling the farmers to increase their income through increased cane production.

Advice on appropriate use of inputs was one of the activities used to measure private extension services. To elicit information on the use of appropriate input, the farmers were asked to respond to statement designed to elicit positive knowledge of performance resulting to use of appropriate input. A 5-point likert scale was constructed to record these responses. Table 6 show results on public extension services in re-lation to appropriate use of input.

The results in table 6 showed that majority of the farmers (75%) receive advice on use of appropriate input from private sector.

A comparison of public and private extension services

Table 7 shows a comparison of the role public and private extension play with respect to various cane farming activities. The relevant information was elicited by asking the farmers to state from whom they receive extension services from, followed by their responses to statements designed to elicit positive knowledge of performance to various cane farming activities. A 5-point likert scale was used to record these responses.

The results in table 7 indicate that, public extension has a lesser role in enhancing effectiveness of outgrowers' credit system as compared to private extension services. This was because the majority of the public extension recipients (61.5%) either disagreed or strongly disagreed compared to 47% private extension recipients in terms of advising farmers on various farm activities.

Table 8: Chi-square test for Effectiveness of Outgrowers Credit System by Type of Extension

	Value	df	Asymp. Significance 2 sided
Pearson Chi-square	14.952	18	0.667
No. valid cases	103		

Significance set at (α = 0.05)

Hypotheses testing

The null hypothesis tested stated that there is no significant difference between the public and private extension services in terms of enhancing effectiveness of out-grower's credit system in Kisumu County. Both the public and private sector extension services were measured with respect to advice given to farmers on land preparation, use of appropriate inputs and creation of awareness. The information on public and private extension services with their effects on effectiveness of out-grower's credit was elicited by use of farmers' questionnaire. Testing of this hypothesis was carried out by use of chi-square test and the results are presented in Table 8.

Results in Table 8 indicate that, there was no significant difference between public and private sector extension services. This was because the Pearson chi-square value (14.952) was not significant at $\alpha = 0.05$ (p>0.05). The null hypothesis was thus accepted. This result suggests that public and private sector extension services are inadequate in terms of quantity despite the fact that cane farmers require it to realize a positive change.

DISCUSSION

Akroyd and Smith (2007) noted that lack of agricultural services has negatively impacted on food production. Consequently, in many parts of less developed countries, agricultural extension services often bypass or do not reach the rural farmers (FAO, 1997). In most countries extension services provided by the government are supplemented by private sectors. Milu and Jayne (2006) acknowledged that, in developing countries, the private sector extension is extremely diverse. Depending on the particular economic and political situation, the private sector may consist of individual farmers/ farm enterprises of all sizes, agricultural input industries, agro-services enterprises, processing industries, marketing farms and multinational firms. It may also include a wide range of agricultural production and marketing co-operatives, farmers associations and private and voluntary organizations. Despite their differences, all these organizations share a common market orientation. They all try to make profit by selling goods and

services. As a result all these private sector organizations have a strong incentive to deliver goods and services (including agricultural extension) efficiently and effectively so as to enhance their ability to survive. Firms that supply agricultural inputs such as seed, chemical fertilizer, pesticides may provide farmers with a wide range of technical and managerial information (through various outreach mechanisms) both to assure that their products are used correctly and also increase agricultural production and income to the farmers. These also motivate customers to buy more products in future (Milu and Jayne 2006). Examples of these private extension agencies are the Muhoroni Sugarcane Outgrowers Company and Chemelil Outgrowers Company, which are currently operating and supporting farmers.

The findings of this study indicated that majority of farmers (82.6%) receive information on credit facility from either from public or private extension providers. The results agree with the findings by Khasiani (1992) who indicated that agricultural technologies might not be adopted if the farmers are not aware of its existence. He continued that lack of awareness acts as a hindrance to the effective participation in agricultural activities. Similarly, Madhur (2000) argued that, impact would be limited if extension is unable to appreciably increase the level of farmers awareness. Further, the results also supports the findings by Mbata (1991) who acknowledged that through extension services the small-scale farmers should be made to understand that credit supervision is for his / her own interest and that, through supervision, credit would be better managed and used for the intended purposes which in turn will increase his productivity and raise their capital base.

The findings of the study also showed that majority of the farmers received extension services on land preparation from private sector. Perhaps this is due to the diverse nature of private sector extension where land preparation machines are readily provided to try and promote return on investment as well as enabling the farmers to increase their income through increased cane production. However, among the farmers who received advice from public extension officers it was noted that the services were not satisfactory. This suggests that perhaps the declining cane production in Kisumu County is due to poor and inadequate land preparation, culminating from inadequate machinery to prepare land for cane growing.

The results further showed that majority of the farmers receive advice on use of appropriate inputs from private sector. This suggests that probably a few farmers may be benefiting as the private sector normally targets potential farmers to maximize the profit from their products. Absence of dependable information to farmers on inputs, on credit and marketing would erode the credibility of extension, hence the rate of adoption by, farmers would be low (FAO, 1994, Khasiani, 1992). However there should be a positive correlation between farmers' link with information sources and adoption (World Bank, 1992 & Chitere, 1995).

This result suggests that public and private sector extension services are inadequate in terms of quantity, that is, in terms of extension agent contact with the farmer despite the fact that cane farmers require it to realize a positive change. Perhaps inadequate extension from government is due to the retrenchment of many staff in the Ministry of Agriculture in the study area (Owuor, 2002).

CONCLUSIONS

Based on the findings of the study a number of conclusions were drawn:

-Public extension service has a role in enhancing effectiveness of the outgrowers' credit system among smallholder cane farmers in Kisumu County. However, the sector needs to enhance the following: provision of advice with respect to, land preparation and use of appropriate inputs.

- That except for inadequate quantity of extension, private extension service plays a greater role in enhancing effectiveness of outgrowers' credit system.

- That in terms of enhancing effectiveness of outgrowers' credits system there was no significant difference between public and private

sector extension services.

Recommendations

From the findings of the study, the following recommendations were suggested.

- Intensification of both public and private extension services.

- Strengthening factory extension division.

- Increasing the number of extension personnel.

- Establishing the contribution of extension among other factors in cane production.

REFERENCES

1- Agbamu, I. U. (2005). 'Problems and Prospects of Agricultural Extension Services in Developing Countries.' In: Adedoyin, S. F. (ed) op cit Pp 159-169.

2- Agribusiness Development Support Project, (2001). Annual Report. Kisumu: Lagrotech Limited.

3- Akroyd, S., & Smith, L. (2007). The Decline in Public Spending to Agriculture – Does it Matter? Briefing Note, No. 2, Oxford Policy Management Institute, Oxford.

4- Anderson, J. R. (2007). Agricultural Advisory Services. Background Paper for the World DevelOpment Report 2008. http://siteresources.worldbank.org/INTWDR2008/Resources/2795087-1191427986785/Anderson AdvisoryServices.pdf.

5- Chitere, P. A. (1995). Extension Education and Farmers Performance in Improved Crop Farming in Kakamega District (Kenya). Agricultural Administration. 18: 39-57.

6- Economic Survey (2007). Central Bureau of Statistics. Ministry of Planning and National Development. Government of Kenya. 2007.

7- Food and Agriculture Organization. (1997). Effectiveness of Agricultural Extension Services in reaching Rural Women in Africa, Volume 2. Italy, Rome: FAO.

8- Gautam, M. & Anderson, J. R. (1999). Reconsidering the Evidence on Returns to T&V Extension in Kenya. Policy Research Working Paper 1098, the World Bank, Washington D. C.

9- Guda, E., Otieno, L.O., Ko'bonyo, P., Okumu, B., Ohito, D., Odera, J., Ogallo, O.S., Rasugu, O., & Odudo, J. (2001). Business and Investment Insight: (Abstract). Maroko Investments Advisory Services Publications.

10- Government of Kenya. (2005). Review of the National Agricultural Extension Policy (NEAP) and its Implementation. Volume II – Main Report and Annexes. Ministry of Agriculture and Ministry of

Livestock and Fisheries Development. Nairobi.

11- KESREF (2007). Kenya Sugar Research Foundation. Strategic Plan 2009-2014.

12- Khasiani, C. (1992). Towards Legitimisation of African Women Indigenous Knowledge in Natural Resource Management. Award News Kenya: Issue Number 4. December, 1993.

13- Kibet, J. K., Omunyinyi, M. E., & Muchiri, J. (2005). Elements of Agricultural Extension Policy in Kenya. Challenges and Opportunities. African Crop Science Conference Proceedings. 7: 1491 - 1494.

14- Koyenikan, M. J. (2008). Issues for Agricultural Extension Policy in Nigeria. International Journal of Agricultural Extension. 12:51-61.

15- Madhur, G. (2000). Agricultural Extension: The Kenya Experience, an Impact Evaluation. Washington D. C: The World Bank.

16- Mbata, J. N. (1991). Agricultural Credit Scheme in Nigeria, A Comparative Study of the Supervised and Non- Supervised Agricultural Credit Scheme as a Tool for Agricultural Development in Rivers State Nigeria. Discovery and Innovation. (Abstract).

17- Milu, M., & Jayne, T.S. (2006). Agricultural Extension in Kenya: Practice and PolicyLessons. Tegemeo Institute of Agricultural Policy and Development, Egerton University.

18- Nambiro, E., Omiti, J., & Mugunieri, L., (2005). Decentralization and Access to Agricultural Extension Services in Kenya. SAGA Working Paper.

19- Ogunsumi, L.O. (2004). Analysis of Sustained Use of Agricultural Technologies on Farmers' Productivity in Southwest, Nigeria. Ph.D. Dissertation, Department of Agricultural Economics and Extension, Federal University of Technology, Akure, Nigeria.

20- Owuor, G. (2002). The Effect of Financial Self-help Groups Credit on Agricultural Production, A Case of Ukwala Division in Siaya District. (Unpublished MSc Thesis), Njoro: Egerton University.

21- Rees, D. M., Wekundah, F., Ndungu, J., Odondi, A.O., Oyure, D., Andima, M., Kamau, J., Ndubi, F., Musembi, Mwaura, L., & Joldersma, R. (2000). Agricultural Knowledge and Information in Kenya-Implications for technology dissemination and development. ODI Agricultural Research and Extension Network Paper (Abstract).

22- Wangia, C. (2001). Micro- Finance Experience in Kenya. In: Anandajayasekeram, Dixon, Kashuliza, Ng'anjo, Tawonezvi, Torkelsson, Wanzira (Eds). Micro – Finance Experience of FARMESA Member Countries in East and Southern Africa. Farmesa, Harare, Zimbabwe.

23- World Bank. (1992). Trends in Agricultural Diversification: Regional Perspectives. Technical Paper No. 180, Washington D.C: World Bank.

24- World Bank (2007). World Development Report 2008: Agriculture for Development, World Bank Washington D.C.

Applying CVM for Economic Valuation of Drinking Water in Iran

Morteza Tahami Pour and Mohammad Kavoosi Kalashami [*]

Keywords:
Economic Value of Water, Willingness to Pay, Municipal Water Consumption, Contingent Valuation Method, Kohkiloye & Boyerahmad Province

Economic valuation of water is useful in the administration and management of water. Population growth and urbanization caused municipal water demand increase in Iran. Limited water resource supply and urban water network capacity raised complexity in water resources management. Present condition suggests using economic value of water as a criterion for allocating policies and feasibility study of urban water supply projects. This study use contingent valuation method for determining economic value of drinking water in Kohkiloye & Boyerahmad province. Required data set were obtained from 177 questionnaires by applying stratified random sampling in 2011 year. From 136 investigated urban households 111 ones are willing to pay more for qualified drinking water. Also, from 41 investigated rural households only 3 ones are willing to pay more for qualified drinking water. Results indicated that economic value of drinking water is 6877 Rial per cubic meter.

[1] Ph.D Student of Agriculture Economics, Department of Agriculture Economic, University of Tehran, Iran.
[*] Corresponding author's email: mkavoosi@ut.ac.ir

INTRODUCTION

Iran is one of the arid and semi-arid countries of the world with average precipitation of 251 mm per year. The total renewable water resources of Iran is 130 billion cubic meters, out of which 92 percentages is used for agriculture, 6 percentages for domestic use and services and 2 percentages for industrial uses (Assadollahi, 2009). Rapid population growth and low irrigation efficiency in agricultural sector of Iran have increased the demand for water resources. Therefore, rational management for water supply and demand and optimum use of the available water resources is necessary. Managing water as an economic good is an important way of achieving efficient and equitable use, and of encouraging conservation and protection of water resources. Water resources provide variety of products and services for human being include food, drinking water, hygiene, entertainment and hydropower. Water demand amount and variety increase beside water supply limitations in municipal regions cause competition among users and impose pressure on drinking water networks. Widespread network of water in cities, high cost of drinking water supply and important role of water in sustainable development raised complexity in water resource management in most cities of Iran. Water allocation and using water economic value as a criterion for allocating is one of the mentioned complexity in water resource management (Turner *et al.,* 2004). The value of water in alternative uses is important for the rational allocation of water as a scarce resource, whether by regulatory or economic means. Water economic value could be used as a proper framework for water pricing and water allocating policies. Also, benefit-cost analysis of water supply projects mainly uses water economic value for economic benefit determination. Most of development documents in Iran insist on estimating economic value of water in different uses and applying this amount as a criterion for allocating. Fourth development program of Iran noted economic value of water should be calculated in water basins. Also, in Fifth development program of Iran using water economic value for allocating and supplying

water resource has been mentioned. Many studies estimate economic value of water in agriculture usage applying parametric and non-parametric methods like production function and mathematical programming (Hussain and Young, 1985, Thompson, 1988, Chakravorty and Roumasset, 1991, He and Tyner, 2004). Against irrigation water, drinking water considered as a final commodity in economical modeling and its economic value determination approach completely different from the situation in which water is a production input like in agriculture and industry (Lehtonen *et al.,* 2003). When water considered as a final commodity, contingent valuation method could properly used for its economic value determination. Many studies use CVM for estimating and calculation economic value of drinking water (Gnedenko and Gorbunova, 1998, Gnedenk *et al.,* 1999, Farlofi *et al.,* 2007, Guha, 2007).

According to the importance of economic water valuation in Kohkiloye and Boyerahmad province, present study estimated economic value of drinking water in municipal and rural districts of mentioned province based on willingness to pay of households for proper quality of drinking water. Estimated WTP for drinking water could be used in benefit-cost analysis of urban water supply projects in Kohkiloye and Boyerahmad province.

MATERIALS AND METHODS

Several approaches applied for economic valuation of water which categorize into two main groups include economic valuation of water as a production input and economic valuation of water as an economic commodity. In agriculture and industry, water used as an input and approaches like production function or mathematical programming applied for economic valuation of water. In municipal usage water treated as a final commodity and water consumers gained utility by using water. Hence, in this case approaches like contingent valuation method, choice modeling and water market transactions have been used for estimating economic value of water (Young, 2005). CVM studies are very popular among mentioned approaches used for economic valuation of drinking water, present study apply CVM for achieving to the foresaid goals.

CVM usually used as a standard and flexible tool for measuring non-use values and non-market use values of natural resources and environment (Hanemann *et al.,* 1991 and Hanemann, 1994). CVM determined individuals WTP in hypothetical scenarios (Lee, 1997). In this approach a hypothetical market for drinking water with proper quality established and some proposed price suggested to individuals for acquiring mentioned commodity. Individual's answer to the proposed prices or bids has been formed based on maximizing utility. As shown below, accepting proposed price means that utility gained by accepting is more than the utility of denying proposed price.

$$U(1, Y - A; S) + \varepsilon_1 \geq U(O, Y; S) + \varepsilon_0 \qquad (1)$$

In above equation U is indirect utility, Y is individual income, A is proposed price or bid and S is social-economical characteristics. Utility difference is as below:

$$\Delta U = U(1, Y - A; S) - U(O, Y; S) + (\varepsilon_1 - \varepsilon_0) \qquad (2)$$

According to the model characteristics logit or probit functional form has been used for estimating valuation function. Considering Pi as the probability of accepting proposed price (A) by individual, logit functional form could be formed as below:

$$P_1 = F_\eta(\Delta U) = \frac{1}{1 + \exp(-\Delta u)} = \frac{1}{1 + \exp\{-(\alpha - \beta A + \gamma Y + \theta S)\}} \qquad (3)$$

In which $F_\eta(\Delta U)$ is cumulative distribution function. θ, γ and β are regression coefficients and it is expected that $\theta > 0$, $\gamma > 0$ and $\beta \leq 0$. After estimating above logit function it is possible to calculate expected WTP using integral.

Logit regression model coefficients have been determined using maximum likelihood estimator (Lehtonen and *et al.,* 2003). Integral of $F_\eta(\Delta U)$ between 0 to infinity has been calculated based on below equation:

$$E(WTP) = \int_0^\infty F_\eta(\Delta U) dA = \int_0^\infty (\frac{1}{1 + \exp\{-\alpha^* + \beta A)\}}) dA \qquad (4)$$

In above equation E(WTP) is expected willingness to pay and α^* is adjusted constant which

is constructed as below:

$$[\alpha^* = (\alpha + \gamma Y + \theta S)] \qquad (5)$$

One of the main advantages of logit estimation is that it is possible to investigate change in variables amount on the probability of accepting proposed price by individual i. The probability of bid acceptance by individual defined as below:

$$P_i = F(X_i^* \lambda) = \frac{1}{1 + \exp^{X_i^* \lambda}} \qquad (6)$$

In which, X_i^* is the vector of variables and λ is the vector of coefficients. In order to evaluate the effect of each variable quantity change on the probability of bid acceptance, the derivation of above equation has been calculated (Maddala, 1991):

$$\frac{\partial P_i}{\partial X_{ik}} = \frac{\exp^{X_i^* \lambda}}{(1 + \exp^{X_i^* \lambda})^2} \lambda_k \qquad (7)$$

This equation generates marginal effect of each variable which is very useful in policy making analysis of model results. Requested data set were obtained through a survey using questionnaires. Mainly, information like education, age, individual satisfaction of drinking water quality and quantity, monthly income and expenditure, water consumption quantity and willingness to pay for drinking water has been asked from each individual by interviewing. Present study used stratified random sampling method for determining the sample size. From 177 sample size 136 and 41 questionnaires gathered from urban and rural districts of Kohkiloye & Boyerahmad province, respectively. Explanatory variables which were used in logit model for estimating economic value of drinking water could be summarized in below table 1.

RESULTS AND DISCUSSION

Investigating sample individuals' satisfaction of drinking water quality showed that 7 percentages of rural individuals and 40 percentages of urban individuals satisfied with present condition of drinking water quality. On the other hand, 93 percentages of investigated rural individuals and 60 percentages of investigated urban

Table 1

Explanatory variables	Description
Region	Dummy variable (1= urban district, 0= rural district)
Water quality	Dummy variable (1=satisfaction from drinking water quality, 0=dissatisfaction from drinking water quality)
Water quantity	Dummy variable (1=satisfaction from drinking water quantity, 0=dissatisfaction from drinking water quantity)
Monthly income	In 1000 Rials
Monthly water consumption	In cubic meter
Family size	-
Proposed price or bid	In 1000 Rials
Family supervisor gender	Dummy variable (1=male, 0=female)
Family supervisor age	-
Family supervisor education	-

individuals dissatisfied with drinking water quality. Reasons of dissatisfaction could be summarized as kidney diseases occurrence, parasite

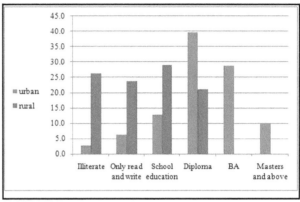

Figure 3: Comparing education level of sample individuals in urban and rural districts (percentages).

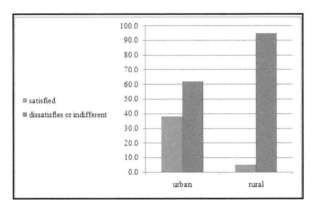

Figure 1: Comparing satisfaction level of drinking water quality in urban and rural districts (percentages).

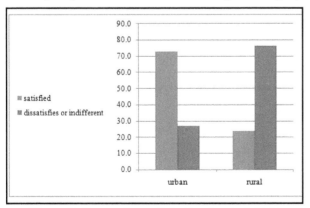

Figure 2: Comparing satisfaction level of drinking water supply in urban and rural districts (percentages).

diseases occurrence and existence of pollution in rural drinking water.

Investigating sample individuals' satisfaction of drinking water quantity showed 17 percentages in rural districts and 69 percentages in urban districts satisfied by present condition of drinking water quantity or supply. Also, 83 percentages of investigated people in rural districts and 31 percentages in urban districts dissatisfied from drinking water quantity or supply. Reasons of dissatisfaction include water quantity, dissection, and network insufficient pressure. Hence, households forced to use tankers or reserve drinking water.

Investigating sample individuals' education level showed that frequency of diploma degree

Table 2: Comparing willingness to pay in urban and rural districts.

District	Description	Willing to pay more	Will not pay more
Urban districts	Frequency	111	25
	percentages	81	19
Rural districts	Frequency	38	3
	percentages	93	7

Source: research findings.

Table 3: Logit model estimation result.

Variables	Coefficients	S.D.	t-statistics	Marginal Effect
Bid	-0.107	0.042	-2.503	-0.024
Monthly income	0.00064	0.00036	1.76	0.00014
Family supervisor education	0.085	0.041	2.061	0.0192
Region	-1.113	0.679	-1.637	-0.25
Constant	1.834	0.639	2.868	-

Log of likelihood function = -86.5
Percentage of Right Predictions = 0.77
Source: Research findings.

is more in urban districts. Education level frequency of sample is shown in figure 3.

From 136 investigated urban households 111 ones are willing to pay more for qualified drinking water. Also, from 41 investigated rural households only 3 ones are willing to pay more for qualified drinking water.

After estimating different models with explanatory variables, below model select as the best estimations.

Results revealed that bid, monthly income, family supervisor education level and dummy of region variables had direct and statistically significant effects on probability of bid acceptance. Marginal effect value of monthly income variable showed that one unit increase (1000 Rials increase) in monthly income 0.0192 unit increase the probability of bid acceptance. Also, marginal effect of bid showed that 1000 Rials increase in bid amount 0.024 unit decrease the probability of bid acceptance by sample individuals. One level increase in family supervisor education 0.0192 unit increase the mentioned probability amount.

Percentages of right prediction in logit model were 77 percentages. So, 77 percentages of individuals' responses could be simulated by model. Using model coefficients, the amount of α^* calculated ($\alpha^* = 2.171$). Expected quantity of WTP for drinking water consumption equals 213187 Rials per month. Considering average drinking water consumption for each household as 31 cubic meters, value of a cubic meter drinking water equals 6877 Rials.

CONCLUSION

Results revealed that drinking water consumers were willing to pay 6877 Rials per cubic meter in Kohkiloye & Boyerahmad province. This price could be used as a good framework for allocating planning of drinking water. Also, for calculating economic benefits of new drinking water supply projects this value could be used. Applying CVM approach for economic valuation of drinking water provides good view for managers and planners about consumers' demand of drinking water.

REFERENCES

1- Assadollahi, S.A. (2009). Groundwater Resources Management in Iran. Secretary General, IRNCID and Deputy of Protection and Exploitation of Iran Water Resources Management Company.

2- Chakravorty, V. and J. Roumasset. (1991). Efficient Spatial Allocation Irrigation Water. American journal of Agricultural Economics. 73: 165-173.

3- Farlofi, S., Mabugu, R., & Ntshingila, S. (2007). Domestic Water Use and Aalues in Swaziland: Contingent Valuation Analysis, Agrekon Magazine. (Abstract)

4- Gnedenk, E., Gorbunova, Z., & Safonov, G. (1999). Contingent Valuation of Drinking Water Quality in Samara City, Moscow State University, zone "I", room 75, Moscow, 117234, Vorobievy Gory.

5- Gnedenko E.D., & Gorbunova, Z.V. (1998). A Contingent Valuation Study of Projects Improving Drinking Water Quality", Modern Toxicological Problems. (Abstract).

6- Guha, S. (2007). Valuation of Clean Water Supply by Willingness to Pay Method in Developing Nation, a Case Study in Calculate, India. Jabalpur University.

7- Hanemann, W. M. (1994). Valuing the Environment through Contingent Valuation. Journal of Economic Perspectives. 8(4): 19-43.

8- Hanemann, W. M., Loonis, J., & Kanninen, B. (1991). Statistical Efficiency of Double-bounded Dichotomous Choice Contingent Valuation. American Journal of Agricultural Economics, 73(4): 1255-1263.

9- He, L., & Tyner, W. (2004). Improving Irrigation

Water Allocation Efficiency Using Alternative Policy option in Egypt, http://econpapers.hhs.se

10- Hussain, R.Z., & Young, R.A. (1985). Estimates of the Economic Value Productivity of Irrigation Water in Pakistan from Farm Surveys, Water Resources Bulletin. 26(6): 1021-1027.

11- Lee, C. (1997). Valuation of Nature-based Tourism Resources Using Dichotomous Choice Contingent Valuation method, Tourism Management. 18(8): 587-591.

12- Lehtonen, E., Kuluvainen, J., Pouta, E., Rekola, M., & Li, C. (2003). Non-market Benefits of Forest Conservation in Southern Finland. Environmental Science and Policy. 6: 195-204.

13- Maddala, G.S. 1991. Introduction of Econometrics. 2nd Edition, Macmillan, New York.

14- Thompson, C.D. (1988). Choice of Flexible Functional Forms: Review and Appraisal, Western Journal of Agricultural Economics. 13: 169-183.

15- Turner, J.C., Douglas, C.L., Hallum, C.R., Krausman, P.R., & Ramey, R.R. (2004). Determination of Critical Habitat for the Endangered Nelson's bighorn sheep in southern California. Wildlife Society Bulletin, 32: 427-448.

16- Young, R.A. (2005). Determining the Economic Value of Water. Concepts and Methods. Washington, DC: Resources for the Future.

8

Perceptions of Constraints Affecting Adoption of Women-in-Agriculture Programme Technologies in the Niger Delta, Nigeria

Iniobong A. Akpabio, Nsikak-Abasi A. Etim[1]* and Sunday Okon

Abstract

The study focused on constraints affecting the adoption of innovative agricultural technologies disseminated by the Women-in-Agriculture (WIA) unit of the Akwa Ibom Agricultural Development Programme (AKADEP) to its women clientele. The study also ascertained the awareness and adoption levels of such introduced technologies. Findings revealed that respondents were aware of 61.9% of introduced technologies, while only 33.3% were fully adopted. The study also identified seven factors responsible for the non-adoption of women farmers' related technologies. The three highest ranking constraining factors were revealed as; high cost of inputs, low income level of women farmers and lack of regular contact with WIA extension agents. Reasons have been proffered for the relatively low technologies' adoption levels. Recommendations have also been made to enhance the technology adoption level. These include the necessity to introduce only socio- economically and culturally compatible technologies to WIA clientele, a wholesale focus on follow-up activities after initial group based technology introduction activities, and the attachment of a credit scheme to the WIA program.

Keywords:
Constraints to Adoption, Awareness, Agricultural Technologies, Women farmers.

[1] *Department of Agricultural Economics and Extension, University of Uyo, PMB 1017, Uyo, Akwa-Ibom State, Nigeria.*
* Corresponding author's email: etimbobo@yahoo.com

INTRODUCTION

Agriculture constitutes a large share of national output and employs a majority of the labour force in most developing countries; hence the sector has been integrated into any thinking about development (World Bank, 2003). However, whereas agriculture-led growth played an important role in slashing poverty and transforming the economies of many Asian and Latin American countries, the same has not occurred in Africa, including Nigeria (Diao *et al.,* 2007). According to Baker (2005), technical change is the engine of long-term growth and it becomes technically important through diffusion. This is more so for agricultural production, where the prospect of enhanced production offered by improved agricultural technologies is recognized, according to the World Food Program, as essential to improving the household food security of small scale farmers, raising rural incomes and creating national surplus that can improve the basis for economic growth (WFP, 1998).

Baker (2005) took a retrospective view at Africa's lack of robust economic growth and dearth of modern technology and concluded that technology (especially agricultural technology) diffusion appears to have failed therein. Eicher (1992) revealed that nearly 100% of the increase in food production in the West African sub-region, since 1960, has come from expanded harvest area, rather than improvements in technology, a trend which Sanders (1996) has deemed, inefficient and with negative long term prospects.

Jafry (2000), Brown *et al.,* (2001) and the Directorate for International Development (DFID, 2004), among many other authors and research scientists, revealed that women are the key farmers, food producers and natural resource managers, in most countries of sub-Sahara Africa. This is because they provide 65 – 89% of food, provide nearly half of farm labour, shoulder over 90% of domestic responsibilities and work twice as many hours as men. Akpabio (2005) also reported an African study which revealed that women carry over 80 tonnes of fuel, water and farm produce for a distance of more than one kilometer over the course of a year.

Despite all these contribution, the Technical Centre for Agricultural and Rural Cooperation, asserted that women are still restricted in their roles as farmers by unequal rights and unequal access to and control over resources, especially land (CTA, 2000). Women also carry out their work without much help from agricultural support mechanisms such as extension agencies, input suppliers and credit institutions (FAO, 2000).

The Women-in-Agriculture (WIA) sub-component of the Agricultural Development Programme (ADPs) was instituted in 1988 to address gender specific agricultural problems. The focus is on food nutrition, processing, storage and utilization of crop and livestock produce, in order to raise women's income and living standards through business oriented farming and processing strategies. Ever since the introduction of the WIA programme in Nigeria, and with the current emphasis on participatory extension, various efforts have been made to elicit various types and levels of information on the activities and effectiveness of the programme in specific limited areas (states) of Nigeria and the Niger Delta. Akpabio (2005b) reported that the WIA programme in Akwa Ibom State remains less than effective, in terms of its contribution to the upliftment of the economic and socio-psychological status of rural women while Adetoun (2000) in South Western Nigeria, and Eshiett (2007) with reference to Akwa Ibom State, revealed that only a few of the technologies disseminated to WIA clientele have been fully adopted.

The importance of agricultural technologies in the development process cannot be overemphasized. It is against this background that this study sought to ascertain clientele perceptions on reasons for the reported low trend of adoption of agricultural technologies. This study however covers the larger South – South (Niger Delta) region of Nigeria, hence it was decided first of all to ascertain on a wider scale of the Niger Delta, the validity of earlier reports of Akpabio (2005b) and Eshiett (2007). In essence, the study sought to answer pertinent questions relating to: (i) the level of women farmers' awareness of specified innovations introduced through the WIA programme and (ii) respondents' perceptions of constraints affecting adoption of technologies disseminated through the WIA programme in the Niger Delta.

MATERIALS AND METHODS
Study Area

The Niger Delta is located in the Southern part of Nigeria. It spreads over a total land mass of about 75,000 square kilometers. It is inhabited by an estimated 30 million population. The people are distributed into forty ethnic groups in about 13,329 communities/settlements in nine states. It is characterized by wetlands and water bodies, with creeks and rivers criss-crossing the entire Southern parts and is often regarded as the largest wetland in Africa and the third largest in the world. The region is however endowed with natural resources. It has the third largest mangrove forest, with the most extensive fresh water swamp forest and tropical rainforest characterized by great biological diversity. Alongside its immense potentials for agricultural revolution, the study area also hosts vast reserves of non-renewable natural resources, particularly hydro-carbon deposits in oil and gas.

The population for the study comprised all the leaders/representatives of different WIA groups who attended the various one-day interactive fora organized by the ADPs in all states in the region. Relevant data could be collated for five states. These were Akwa Ibom, Cross River, Delta, Edo, and Rivers. All the 267 participants were purposefully utilized for the study, although responses from 250 respondents were eventually utilized for data analysis (viz, table 1). A pre-tested and validated structured Interview Schedule and Focus Group Discussions were utilized to elicit relevant information from the selected sample. These activities were performed with the aid of trained enumerators.

To ascertain the level of women farmers' awareness and adoption of specified innovations introduced through the WIA program, a list of technologies disseminated through the WIA programme was obtained, after which awareness and adoption scores were computed for each technology. Scores of 0 and 1 were recorded for awareness and non-awareness of disseminated technologies while scores of 2, 1 and 0 were recorded for adopted, discontinued and non-adopted technologies, respectively. A mean cut-off score of 0.5 was adopted to demarcate between technologies for which respondents were either 'aware' or 'not aware', while a cut-off mean score of 1.0 was utilized to differentiate between technologies which have either been 'adopted' or 'not adopted'. In essence, respondents were deemed to be aware of a technology with a mean score of 0.5 and above, while they were not aware of technologies with mean scores of less than 0.5. Similarly, technologies which recorded mean scores of 1.0 and above were perceived as adopted by respondents, unlike technologies with mean scores of less than 1.0, which were regarded as not adopted.

To determine respondents' perceptions of constraints affecting adoption of WIA programme technologies, a list of possible constraints that may hinder the adoption of disseminated technologies was drawn up with the aid of interviews and literature search. A 3-point Likert continuum of agreed (3) undecided (2) and disagreed (1) was employed to compute responses on reasons for non-adoption of WIA technologies. A cut-off mean score of 2.5(3+2+1/3 +0.5) was utilized to differentiate between 'major' and 'minor' factors for non –adoption, where a score of 2.5 and above, was depicted as a 'major' factor for non-adoption, while items with scores below 2.5 were adjudged minor factors.

RESULTS AND DISCUSSION
Awareness and adoption levels of WIA technologies

Tables 2 and 3 show that women farmers were aware of 61.9% (13 of 21) introduced

Table 1: Selected Sample

S/N	State	Population	Sample
1	Akwa Ibom	51	51
2	Cross River	60	53
3	Delta	53	48
4	Edo	56	54
5	Rivers	47	44
	TOTAL	**267**	**250**

Table 2: Distribution of Respondents based on the extent of Awareness of WIA technologies

	AKADEP Technologies Food Crops	Aware (1)	Not aware (0)	Means	Remarks
1	Cassava/maize/melon planting	230 (92) *	20 (8)	0.92	Aware
2	Yam mini set	124 (49.6)	126 (50.4)	0.49	Not Aware
3	Dry season vegetable	170 (68)	80 (32)	0.68	Aware
4	Wet season vegetable	210 (84)	40 (16)	0.84	Aware
5	Rice cultivation	86 (34.4)	164 (65.6)	0.34	Not Aware
	Processing & Utilization				
6	Soya bean milk/ flour	140 (56)	110 (44)	0.56	Aware
7	Odorless fufu/garri	250 (100)	0 (0)	1.00	Aware
8	Fruit drinks	116 (46.4)	134 (53.6)	0.46	Not Aware
9	Plantain chips processing	210 (84)	40 (16)	0.84	Aware
10	Pineapple chips processing				
	Input use	200 (80)	50 (20)	0.80	Aware
11	Fertilizer use	180(72)	70(28)	0.72	Aware
12	Improved crop varieties e.g. maize	160 (64)	90 (36)	0.64	Aware
13	Agro chemicals e.g. Pesticides	84 (33.6)	166 (66.4)	0.33	Not Aware
14	Improved animal breeds	130 (52)	120 (48)	0.52	
	Agroforestry technology				
15	Snail rearing	124 (49.6)	126 (50.4)	0.49	Aware
16	Plantain /Cocoyam intercropping	196 (78.4)	54 (21.6)	0.78	Aware
17	Afang cultivation	210 (84)	40 (16)	0.84	Aware
18	Bee raising	78 (31.2)	172 (68.8)	0.31	Aware
	Tree crops planting				
19	Improved oil palm seedlings	160 (64)	90 (36)	0.64	Aware
20	Rubber seedlings	90(36)	160 (64)	0.36	Not Aware
21	Improved cocoa seedlings	70 (28)	180(72)	0.28	Not Aware

*-Percentages in parentheses

technologies, while only 33.3% (7 of 21) of the technologies were eventually adopted. It was also observed on table 3, that respondents adopted 53.9% (7 of 13) of the technologies for which they were aware. A related finding in the course of study revealed that only 59.2% respondents received information on improved agricultural technologies from extension officials of the WIA program, while 20.8% and 20% respondents received information from relatives/friends and husbands, respectively. There is a cause for concern here. This is because an extension program deliberately targeted at women farmers reaches only 59.2% of intended clientele. Many reasons have been proffered for this undesirable situation. These include; long distance from meeting venues and concomitant non-attendance at group meetings (Adetoun, 2000) lack of interpersonal contact, arising from lack of follow-up after group meetings (Udoh, 2001) and lack of relevance of disseminated messages to the amelioration of female farmers livelihood constraints (Reij and Waters-Bayer, 2001) among many others, might

have led to clientele' loss of interest in extension offerings.

Table 3 also shows that none of the technologies disseminated to respondents under "input use" and "tree crops planting" classifications was adopted. Odourless fufu/garri (fresh/dried cassava paste) x = 1.96; cassava/maize/melon crops combination (x = 1.76) and intercropping (x = 1.45) were the most adopted technologies. This result corroborates Baker's (2005) and Swinkels and Franzel's (1997) assertion that compatible technologies and technologies that differed very little from the old technologies would diffuse faster since there would be less of an information problem associated with them. Pannell (1999) described four conditions necessary for farmers to adopt innovative technologies, two of which are "awareness of the technology" and "perception that technology promotes farmers objectives". It may be inferred that farmers will adopt more of the technologies for which they are aware. In essence, awareness of technology is a motivating factor for the adoption of technological packages. Hardarker, Huirne and Anderson (1997) however

Table 3: Distribution of respondents based on the extent of adoption of WIA technologies.

AKADEP Technologies (Food Crops)	Not Adopted (0)	Discontinued (1)	Adopted (2)	Means**	Remarks
1 Cassava/maize/melon planting	0 (0) *	20 (8)	210 (84)	1.76	Adopted
2 Yam mini set	0 (0)	88 (35.2)	36 (14.4)	0.64	Not Adopted
3 Dry season vegetable	60 (24)	60 (24)	48 (19.2)	0.62	Not Adopted
4 Wet season vegetable	0 (0)	40 (16)	170(68)	1.52	Adopted
5 Rice cultivation	172 (34.4)	0 (0)	0 (0)	0.00	Not Adopted
Processing & Utilization					
6 Soya bean milk/ flour	10 (4)	70 (28)	60 (24)	0.76	Not Adopted
7 Odourless fufu/garri	0 (0)	10 (4)	240 (96)	1.96	Adopted
8 Fruit drinks	110 (44)	6 (2.4)	0 (0)	0.02	Not Adopted
9 Plantain chips processing	4 (1.6)	16 (6.4)	190 (76)	1.58	Adopted
10 Pineapple chips processing	14 (5.6)	10 (4)	176 (70.4)	1.45	Adopted
Input Use					
11 Fertilizer use	40 (16)	52 (20.8)	88 (35.2)	0.91	Not Adopted
12 Improved crop varieties e.g maize	24 (9.6)	60 (24)	76 (30.4)	0.84	Not Adopted
13 Agro chemicals eg. pesticides	34 (13.6)	46 (18.4)	4 (1.6)	0.21	Not Adopted
14 Improved animal breeds	40 (16)	20 (8)	60 (24)	0.56	Not Adopted
Agroforestry technology					
15 Snail rearing	16.(6.4)	24 (9.6)	84 (33.6)	0.76	Not Adopted
16 Plantain /Cocoyam intercropping	12 (4.8)	4 (1.6)	180 (72)	1.45	Adopted
17 Afang cultivation	30 (12)	16 (6.4)	164 (65.6)	1.36	Adopted
18 Bee raising	78 (31.2)	0 (0)	0 (0)	0.00	Not Adopted
Tree crops planting					
19 Improved oil palm seedlings	140 (56)	20 (8)	0 (0)	0.08	Not Adopted
20 rubber seedlings	82 (32.8)	8 (3.2)	0 (0)	0.03	Not Adopted
21 Improved cocoa seedlings	66 (26.4)	4 (1.6)	0 (0)	0.01	Not Adopted

*-Percentages in parentheses
** Mean Scores calculated, based on total no. of respondents- regardless of the no. of actual recorded responses per innovation.

cautioned that high level awareness of technologies does not necessarily translate into higher adoption levels. This is because farmers will only adopt those innovations which are adjudged useful and beneficial to their particular situation. A disaggregated analysis of table 2 reveals relatively high frequencies of "non-awareness" scores that were recorded for some technological offerings which were generally perceived (mean scores) as being "aware" of by respondents (viz; items 3, 6, 12, 14 and 19). Bunch (1982) and Baker (2005) harped on the importance of critical mass in the adoption of innovations. The researchers contended that that there is a higher level of adoption and less discontinuance for a new technology in which the whole community or a critical mass (proportionately larger than average) of farmers are aware of and in which they are interested. They explain reasons for this in terms of traditional communities being accustomed to living in an environment of consensus, and that schemes which entail community risk sharing are more easily imbibed than otherwise.

Constraints affecting adoption of WIA programme technologies

Results as shown on table 3 reveals that respondents perceived 7 of the 21 identified items as possible reasons for the relatively low adoption levels of agricultural technologies introduced through the WIA programme. These are: high cost of inputs (x = 3.0), low income level of women farmers' (x = 2.97) lack of regular contact with extension agents (x =2.82) old age of women farmers in the study area (x = 2.73) poor attitude towards risk and change (x = 2.55) and complexity of introduced technologies (x = 2.55). The above revealed findings find relevance in related literature. Rogers (1995) identified five key characteristics of innovations that determine their adoption potential, including: relative advantage, trialability, compatibility, observability and complexity. Reed (2001) iden-

Table4: Distribution of respondents based on perception of factors affecting non-adoption of WIA technologies.

	Reasons for Non Adoption/ Discontinuace	Disagreed (1)	Undecided (2)	Agreed (3)	Mean (X)	Remarks
1	High cost of inputs	0(0) *	0(0)	250 (100)	3.00	Major Factor
2	Lack of supporting inputs	88(35.2)	62(24.8)	100(40)	2.05	Minor Factor
3	Problem of diseases / pests	66 (26.4)	34(13.60)	150 (60)	2.33	Minor Factor
4	Non-appropriateness of the technological package to the Local environment	60 (24.1)	68 (27.2)	122 (48.8)	2.25	Minor Factor
5	Non-availability of the improved package	84 (32.80)	16 (6.4)	152 (60.8)	2.28	Minor Factor
6	Non-Profitability of the new technology	188 (75.2)	40 (16)	22 (8.8)	1.16	Minor Factor
7	Superiority of the old technology to the newly introduced one.	174 (69.6)	48 (19.2)	28 (11.2)	1.42	Minor Factor
8	Incompatibility of the new technology with the norms and customs of the local environment	144 (57.6)	60(24)	46 (18.4)	1.60	Minor Factor
9	Lack of clear understanding of the newly introduced package	12(4.8)	104(14.6)	134 (53.6)	2.48	Minor Factor
10	Low level of educational attainment by women farmers in the study area.	48 (19.2)	106 (42.4)	96(38.4)	2.19	Minor Factor
11	Low level of income of women farmers in the area	2 (0.8)	4 (1.6)	244 (97.6)	2.97	Major Factor
12	Insufficient Programs designed to convince and encourage farmers to change	12 (4.8)	144 (59.6)	94 (37.6)	2.32	Minor Factor
13	Women farmers perception of the old technology as better than the new one.	138 (55.2)	68 (29.2)	44 (17.6)	1.62	Minor Factor
14	Inconsistence of the innovation with the existing farming system, values and needs of women farmers in the area	156 (62.4)	36(14.4)	58 (23.2)	1.61	Minor Factor
15	Inadequate information about the newly introduced technological package.	50(20)	58 (23.2)	142(59.2)	2.44	Minor Factor
16	Complexity of the introduced innovation.	44(17.6)	24 (9.6)	182 (92.8)	2.55	Major Factor
17	Failure of some demonstration plots set –up by the extension agents.	138 (55.2)	58(23.2)	54 (21.6)	1.66	MinorFactor
18	Lack of regular contact with extension agents	20(8)	4(1.6)	226(90.4)	2.82	Major Factor
19	Poor attitude of women farmers towards change and risk	8 (3.2)	60 (24)	182 (72.8)	2.70	Major Factor
20	Age of women farmers in the study area	20 (8)	28 (11.2)	202 (80.8)	2.73	Major Factor
21	Lack of access and control over production resources such as land and credit facilities.	8(3.2)	96(38.4)	146(58.4)	2.55	Major Factor

*-Percentages in parentheses

tified the most significant of these characteristics, as: high relative advantage, high compatibility and low complexity. Swinkels and Franzel (1997) agreed with the submission above, but also opined that for the female gender, additional incentives for adoption may include factors like, suitability to accepted gender roles, cultural acceptance and compatibility with other enterprises.

High cost of inputs for introduced technologies and low income of respondents' were revealed as the greatest constraints to adoption of introduced technologies. Obinne (1994) and Arokoyo (1996) mentioned low income level of farmers and high cost of inputs as constraints to technology adoption, especially among low income

farmers. In that wise, Baker (2005) and Hebinck, Franzel and Richards (2007) asserted that the most successful programmes of agricultural change are those that tie adoption to credit programmes. Udoh (2001) and Eshiett (2007) maintained that contact with extension agents, especially with respect to interpersonal contacts, relate favorably to the adoption of new farm practices and concomitant improved agricultural production. Obinne (1994) and Baker (2005) opined that poor attitude to risk, in terms of excessive risk aversion may severely limit adoption of technological innovations especially among female rural farmers, while Baker (2005) opined that technologies that differed very little from the old technologies would diffuse faster than unrelated technologies, while the older generation may credibly block adoption even if the younger generation co-ordinates. Dove (1991) contended that individuals with insecure tenure will generally be less likely to invest in new technologies that require complementary immobile inputs, while Due, Mudenda and Miller (1993) asserted that although women want to increase the productivity of the resources they control, they face greater obstacles to change. One of such obstacles, according to Reij and Waters-Bayer (2001) is lack of relevance of disseminated messages to the amelioration of female farmers' livelihood constraints. (PLEASE, THIS IS NOT LITT REVIEW. THIS IS SIMPLY A COMPARISON OF FINDINGS WITH RESULTS FROM PREVIOUS STUDIES)

It is obvious that although poor female farmers in the study area are conscious of innovating in order to overcome their present precarious socio-economic situation, they are however precluded from benefiting from opportunities open to them due to various constraining factors, as have been identified above.

CONCLUSION AND RECOMMENDATIONS

It has been revealed that the WIA program, as being implemented in Akwa Ibom State does not reach out to a large number of its intended clientele base. This has resulted in an average level of awareness and concomitant relatively lower level of adoption of innovative technologies disseminated by WIA extension officials. The study also identified seven factors which combine to hinder the adoption of disseminated WIA technologies. The major constraints were: high cost of inputs, low income level of women farmers and lack of regular contact with WIA extension agents. Many reasons, backed by literature, have been proffered for this trend, including the fact that only 59.2 percent respondents regarded WIA extension officials as their source of information on innovative agricultural technologies. In order to enhance the success of the WIA programme in Akwa Ibom State, attempts should be made to ameliorate constraints which hinder extension officials' access to their potential clientele. To aid in this direction, adequate logistic support should be

provided to WIA extension agents so as to help enhance the process of contacting their expectant clientele Technologies slated for dissemination should be compatible to clientele socio-economic and cultural base and emphasis should be focused on follow-up activities, after initial group meetings. This would help to practicalize disseminated technologies on the farms and in the homes of potential adopters of technological innovations. It may also be necessary to attach credit schemes to the WIA program, in terms of linking the various women groups to various credit agencies.

REFERENCES

1- Adetoun, B. A. (2000) Organization and Management of Agricultural Extension Services for Women Farmers in South-Western Nigeria. Centre for Sustainable Development and Gender Issues. Abner Publishing, Ibadan, (Abstract).

2- Akpabio, I.A (1997) Extension Communication Methods as a Strategy for Agricultural Development in Akadep In: Issues in Sustainable Agricultural Development. Fabiyi, Y.L. and M.G. Nyienakuna (eds). Vol. 1. August. Dorand Publishers. Uyo. P. 122.

3- Akpabio, I.A. (2005). Human Agriculture: Social Themes in Agricultural Development. Abaam Pub. Co. Uyo. 219 pp.

4- Akpabio, I.A. (2005b). 'Beneficiary Perceptions of the Gender Specific Extension Delivery Service in Akwa-Ibom, Nigeria.' Gender and Behaviour 396-405.

5- Anderson, J. R and P. B. Hazell (1989). Variability in Grain Yields: Implications for Agricultural Research and Policy in Developing Countries. Baltimore: John Hopkins University Press, p. 23

6- Arokoyo, T. (1996). Agricultural Technology Development and Dissemination. A Case Study of Ghana and Nigeria. CTA. The Netherlands, p. 5.

7- Baker, E. (2005) Institutional Barriers to Technology Diffusion in Rural Africa. University of Massachusetts, Amherst. 30pp.

8- Bunch, R. (1982). Two Ears of a Corn. USA: World Neighbours, p. 11.

9- Brown, L., Feldstein, H., Haddad, L.C. & Pena, A. Quisumbing (2001). Generating Food Security in the year 2020: Women as Producers, Gatekeepers and Shock Absorbers" (In) The Unfinished Agenda. Pinstrup-Andersen and Pandya-Lorch (eds) IFPRI. Washington D.C. pp. 205-209.

10- Diao, X., P. Hazell, D. Resnick and J. Thurlow (2007). The Role of Agriculture in Development: Implications for SSA. Research Paper 153. IFPRI. Washington DC, p.5

11- Department for International Development (DFID 2004) "Gender" Fact Sheet. Sept. London.

12- Dove, M. R. (1991) Forester's Beliefs about Farmers: An Agenda for Social Science Research. In: Social Forestry- EAPI Working Paper (Abstract).

13- Due, J., Mudenda, T., & Miller, C. (1983). How Do Rural Women Perceive Development? A Case Study in Zambia. Report no 83-E-265. University of Illinois. Dept of Agricultural Economics, Urbana, IL.

14- Eicher, C. (1992) "African Agricultural Development Strategies. In: Alternative Development Strategies in sub-Saharan Africa. Stewart, F., L. Sanjaya and S. Wengwen (eds) Macmillan. London.

15- Eshiet, E. (2007) "Level of Participation of Women Farmers in the Women in Agriculture Programme in Akwa Ibom State". Unpublished M. Sc thesis. Dept of Agricultural Economics and Extension, Universilty of Uyo, Nigeria. 88 pp.

16- FAO, (2000). Rural Women and Development Agencies. Rome. P. 23

17- Hardarker J. B., Huirne R.B.M & Anderson, J.R. (1997). Coping with Risks in Agriculture. CAB. Int. Oxon. UK.

18- Hebinck, P., Franzel, S., & Richards, P. (2007). Adopters, Testers or Pseudo Adopters? Dynamics of the Use of Improved Tree Fallow by Farmers in Western Kenya. AgroSystems 94(2): 509-515.

19- Jafry, T. (2000). Women, Human Capital and Livelihood – an Ergonomics Perspective. 0D1 Natural Resources Perspectives (Abstract).

20- Pannell, D.J. (1999). Social and Economic Challenges in the Development of Complex Farming Systems. Agroforestry Systems 45: 393-409.

21- Obinne, C. 1994). Fundamentals of Agricultural Extension. ABIC Publishers, Enugu, Nigeria.

22- Reid, M.S. (2001). Paticipatory Technology Development for Agro-forestry: An Innovation – Decision Approach. Centre for Enviornmental Management Working Paper R01/121. University of Leeds.

23- Reij, C., & Waters-Bayer, A. (2001). An Initial Analysis of Farmers. Innovations in Africa. in: A Source of Inspiration for Agricultural Development. Reij and Water Bayer eds. Earthscan Publications pp. 3-22.

24- Rogers, E. M. (1995). Diffusion of Innovations, 4th Edition. The Free Press, NewYork.

25- Swinkels, R., & Franzel, S. (1997). Adoption Potential of Hedgerow Intercropping in Maize Based Cropping Systems in the Highlands of Western Kenya. Economic and farmers' evaluation. Experimental Agriculture 33:211-223.

26- Technical Centre for Agricultural and Rural Co-operation (CTA-2000) The Economic Role of Women in Agricultural and Rural Development: Promoting Income-Generating Activities. C.T.A. Wageningen. The Netherlands, 56pp.

27- Udoh, A.J. (2001). Agricultural Extension Development Administration. Etofia Media Services Ltd. Uyo. p. 39.

28- Udokah, E. (2004). Information Source Preference of Small Scale Farmers in Akwa Ibom State. Un-Published M. Sc thesis. Dept of Agricultural Economics and Extension, University of Uyo, Nigeria, 71pp.

29- World Bank (2003). Sudan Country Economic Memorandum: Stabilisation and Reconstruction. Washington DC. World Bank, p. 3.

30- World Food Programme (1998). Food Aid to Support Technology Adoption Among Small Scale Agriculturists. WFP, Rome, Italy.

9

A Logistic Regression Analysis: Agro-Technical Factors Impressible from Fish Farming in Rice Fields, North of Iran

Seyyed Ali Noorhosseini-Niyaki [1*], Mohammad Sadegh Allahyari [2]

Abstract

This study was carried out to identify Technical-Agronomic Factors Impressible from Fish Farming in Rice Fields. This investigation carried out by descriptive survey during July-August 2009. Studied cities including Talesh, Rezvanshahr and Masal set in Tavalesh region near to Caspian Sea, North of Iran. The questionnaire validity and reliability were determined to enhance the dependability of the results. Data were collected from 184 respondents (61 adopters and 123 non-adopters) randomly sampled from selected villages and analyzed using logistic regression analysis. Results showed that there was a significant positive relationship ($p < 0.05$) between biological control of pests in rice fields and the fish farming in rice fields. Also, there was a significant negative relationship ($p < 0.10$) between the fish farming in rice fields and variables of quantity using pesticide of Diazinon in rice fields and number of plows in rice fields.

Keywords:
Rice-Fish Farming, Technical-Agronomic Factors, Pest, Weed, Plow, Fertilizer

[1]*Department of Agronomy, Lahidjan Branch, Islamic Azad University, Lahidjan, Iran*
[2]*Department of Agricultural Management, Rasht Branch, Islamic Azad University, Rasht, Iran*
* Corresponding author's email: Noorhosseini.SA@gmail.com

INTRODUCTION

The earliest records of fish culture in rice-fields originate from China, circa 2000 years ago, followed by India, 1500 years ago. Other countries with a recorded history of rice-fish culture are Indonesia, Malaysia, Thailand, Japan, Madagascar, Italy, Russia, Vietnam (Rothuis, 1998), Egypt, Philippines, Bangladesh, Cambodia, Korea and other countries (Saikia and Das, 2008; Frei and Becker, 2005; Halwart, 1998). Also in northern Iran rice-fish farming is a new farming system (Karami et al., 2006; Noorhosseini and Allahyari, 2010). Integrated rice-fish farming offers a solution to economic problem of farmers by contributing to food, income and nutrition. Not only the adequate supply of carbohydrate, but also the supply of animal protein is significant through rice-fish farming. Fish, particularly small fish, are rich in micronutrients and vitamins, and thus human nutrition can be greatly improved through fish consumption (Larsen et al., 2000; Ahmed and Garnett, 2011; Noorhosseini and Mohammadi, 2010). Many reports suggest that integrated rice-fish farming is ecologically sound because fish improve soil fertility by increasing the availability of nitrogen and phosphorus (Giap et al., 2005; Dugan et al., 2006; Noorhosseini, 2012).The feeding behavior of fish in rice fields causes aeration of the water. Integrated rice-fish farming is also being regarded as an important element of integrated pest management (IPM) in rice crops (Berg, 2001; Hilbrands and Yzerman, 2004). At the farm level rice-fish integration reduces use of fertilizer, pesticides and herbicides in the field. Such reduction of costs lowers farmer's economic load and increases their additional income from fish sale (Noorhosseini, 2010; Noorhosseini and Radjabi, 2010). Also, integrated rice-fish farming gave higher rice yields and fetched higher gross margin than sole rice cropping system (Das et al., 2002; Hossain et al., 2005). Ahmed and Garnett (2011) Reported that higher yields can be achieved by increasing inputs in the integrated farming system. Integrated rice-fish farming also provides various socioeconomic and environmental benefits. Nevertheless, only a small number of farmers are involved in integrated rice-fish farming due to a lack of technical knowledge, and an aversion to the risks associated with flood and drought. In addition, Ahmed et al., (2011) Reported that rice-fish farming is as production efficient as rice monoculture and that integrated performs better in terms of cost and technical efficiency compared with alternate rice-fish farming. However, a lack of technical knowledge of farmers, high production costs and risks associated with flood and drought are inhibiting more widespread adoption of the practice. Our objective was to identify technical-agronomic factors impressible from fish farming in rice fields in north of Iran.

MATERIALS AND METHODS

Studied Location and Survey: This study was carried out by survey during July and August 2009. Studied area including Talesh, Rezvanshahr and Masal set in Tavalesh region of Guilan province near to Caspian Sea, north of Iran (Figure 1). Respondents were selected from rural area and categorized into adopters and non-adopters of integrated rice-fish farming. Totally 184 farmers were selected by stratified random sampling technique using the table for determining the sample from given population developed by Bartlett et al., (2001) that including 61 (33.15%) adopters and 123 (66.85%) non-adopters (Table 1). This survey was conducted in using a questionnaire with open-ended questions. The questionnaire was pre-tested by in-

Table 1: Total sample size used in the study area

	Talesh	Masal	Rezvanshahr	Total
IRFF Adopters Population	31	31	17	17
IRFF Adopters Sample Size	19	28	14	14
IRFF Non-adopters Sample Size	38	56	29	29

Source: Survey Results, 2009

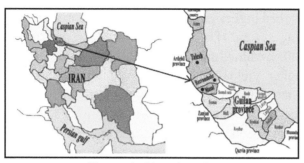

Figure 1: Site of study

terviewing three farmers (not included in the study). After some modifications, it was tested again with 10 other respondents.

Statistical Analysis: The technical-agronomic variables for the two groups were examined using logistic regression model. The dependent variable was the adoption of rice-fish farming. The dependent variable was dichotomized with a value 1 if a farmer was an adopter of integrated rice-fish farming and 0 if non-adopter. The definitions and measurement of variables are present in Table 2. AF, AH, MW, AP, BP and WI were entered in the model as dummy variables. The other variables namely QF, QD and NP were entered as continuous variables. Data analysis was conducted with Statistical Package for Social Sciences (SPSS 18).

The model was specified as follows;

$$Y = f(AF, QF, AH, MW, AP, BP, QD, NP, WI)$$

strength of the joint effect of the covariates on probability of adoption among farmers in the zone. The results also showed that the decision on adoption of rice-fish farming is determined by biological control of pests in rice fields (BP), quantity using Diazinon in rice fields (QD) and number of plows in rice fields (NP) which have significant influence. Also, the Wald indicating the relative contribution of individual variable to probability of adoption of rice-fish farming showed that BP (4.538) was the most important factor determining choice of adoption of rice-fish farming among the rice farmers. Generally, the results of logistic regression show that there was a significant positive relationship ($p < 0.05$) between biological control of pests in rice fields and the fish farming in rice fields (Table 3). These results were similar to Frei and Becker (2005) and Kathiresan (2007). In other words, fish farming in rice fields reduced the use of chemical control methods that is light reason of rice-fish farming system sustainability. Also, there was a significant negative relationship ($p < 0.10$) between the fish farming in rice fields and quantity using Diazinon in rice fields (Table 3). These results are consistent with Saikia and Das (2008), Salehi and Momen Nia (2006). The use of chemical control methods reduced with adoption of rice-fish farming which is also

Table 2: Definition of variables included in the regression model

Dependent variable Y = Adoption	Adopters = 1, Non adopters = 0
Independent variable	Yes = 1, No = 0
AF = Application of Chemical Fertilizers	Kg/ha
QF = Quantity Using Chemical Fertilizers	Yes = 1, No = 0
AH = Application of Herbicides	Yes = 1, No = 0
MW = Mechanical Control of Weed	Yes = 1, No = 0
AP = Application of Pesticides	Yes = 1, No = 0
BP = Biological Control of Pests	Kg/ha
QD = Quantity Using Diazinon	Number in year
NP = Number of Plows	Very mach = 5, Much = 4, Intermediate = 3, Little = 2, Very
WI = Accessibility to Water Supply for Irrigation	little = 1

RESULTS AND DISCUSSION

The results of the Logit likelihood regression model indicated that the overall predictive power of the model (70.1%) is quite high, while the significant Chi square ($p < 0.01$) is indicative of

compatible with sustainable agriculture. Since many fish species feed partly on the aquatic fauna, it has been assumed that they can act as biological control agents in rice fields. Concurrent rice and fish culture decrease pesticides appli-

Table 3: Logistic regression coefficients of the technical-agronomic factors affecting adoption of rice-fish farming

	B	S.E.	Wald	Sig.
AF	-19.131	25320.477	0.000	0.999
QF	0.000	0.001	0.005	0.943
AH	-21.476	18858.287	0.000	0.999
MW	20.493	40192.879	0.000	1.000
AP	0.611	0.558	1.200	0.273
BP	0.996	0.468	4.538	0.033**
QD	-0.029	0.016	3.311	0.069*
NP	-0.717	0.396	3.273	0.070*
WI	0.112	0.167	0.450	0.502
Constant	20.822	51110.003	0.000	1.000

*** $p<0.01$, ** $p<0.05$ and * $p<0.10$
-2 log likelihood = 204.830
Chi square statistic = 28.943***
Overall Correct predictions = 70.1%

cation compare monoculture (Berg, 2002; Noorhosseini, 2011). In general, common carp, being an omnivorous feeder, seems to be the most promising species in controlling insects and snails [15]. In addition, there was a significant negative relationship ($p<0.10$) between fish farming in rice fields and number of plows in rice fields (Table 3). According to results, by adoption of rice-fish farming, number of plows was reduced. It seems that the decrease of tillage frequency among adopters, caused by farms occupied by fishes is the time dimension.

CONCLUSION

In general, our results suggest that biological control of pests in rice fields, quantity using Diazinon in rice fields, and numbers of plows in rice fields were the most important technical-agronomic factors impressible from fish farming in rice fields. In other words, adopters of rice-fish farming used less chemical materials in order to control pests and reduce the number of plows. Furthermore, society health and environment sustainability will be saved and they reach more profit that is economical. Also, since aquaculture requires resources such as pond, land, water and other inputs, poor farmers cannot afford the requirements. As a target to understand and meet their needs and to access the common water resources available in their rice-fields, rice-fish farming is the most appropriate technology in recent times.

REFERENCES

1- Ahmed, N. & Garnett, T.S. (2011). Integrated Rice-fish Farming in Bangladesh: Meeting the challenges of food security. Food Security. 3(1): 81-92.

2- Ahmed, N., Zander, K.K., & Garnett, S.T. (2011). Socioeconomic Aspects of Rice-fish Farming in Bangladesh: Opportunities, Challenges and Production Efficiency. The Australian Journal of Agricultural and Resource Economics, 55: 199–219.

3- Bartlett, J.E., J.W. Kotrlik and C.C. Higgins, (2001) Organizational Research: Determining Appropriate Sample Size in Survey Research. Information Technology, Learning and Performance Journal, 19(1): 43-50.

4- Berg, H. (2001) Pesticide Use in Rice and Rice-Fish Farms in the Mekong Delta, Vietnam. Crop Protection, 20: 897–905.

5- Berg, H. (2002) Rice Monoculture and Integrated Rice–Fish Farming in the Mekong Delta, Vietnam-economic and ecological considerations. Ecological Economics, 41: 95– 107.

6- Das, D.R., Quddus, M.A., Khan, A.H. & Nur-e-Elahi, M. (2002). Farmers Participatory Productivity Evaluation of Integrated Rice and Fish Systems in Transplanted Aman rice. Pakistan Journal Agronomi. 1(2-3): 105-106.

7- Dugan, P., Dey, M.M., & Sugunan, V.V. (2006). Fisheries and Water Productivity in Tropical River Basins: Enhancing Food Security and Livelihoods by Managing Water for Fish. Agricultural Water Management, 80: 262–275.

8- Frei. M., & K., Becker. (2005). Integrated Rice-Fish Culture: Coupled Production Saves Resources. Natural Resources Forum. 29: 135–143.

9- Giap, D.H., Yi, Y., & Lin, C.K. (2005). Effects of

Different Fertilization and Feeding Regimes on the Production of Integrated Farming of Rice and Prawn Macrobrachium Rosenbergii (De Man). Aquaculture Research, 36: 292–299.

10- Halwart, M. (1998). Trends in Rice-Fish Farming. FAO Aquaculture Newsletter, 18: 3-11.

11- Hilbrands, A., & Yzerman, C. (2004). On-Farm Fish Culture. Printed by: Digigrafi, Wageningen, the Netherlands. Second edition. 67p.

12- Hossain, S.T., Sugimoto, H., Ahmed, G.J.U., & Islam, M.R. (2005). Effect of Integrated Rice-Duck Farming on Rice Yield, Farm Productivity, and Rice-Provisioning Ability of Farmers. Asian Journal of Agriculture and Development. 2 (1): 79-86.

13- Karami E. A., Rezaei Moghadam K., Ahmadvand M., & Lari, M.B. (2006). Adoption of Rice- Fish Farming (RFF) in Fars province. Iranian Journal Agriculture. Extension & Education. 2(2): 31-43.

14- Larsen, T., Thilsted, S.H., Kongsbak, K., & Hansen, M. (2000). Whole Small Fish as a Rich Calcium Source. British Journal of Nutrition. 83: 191–196.

15- Noorhosseini-Niyaki, S.A. (2010). Fish Farming in Rice Fields Toward Sustainable Agriculture: The Case Study in Guilan Province. The First National Conference on Sustainable and Cleaner Product, Esfahan of Iran. Nov 10-11, 2010.

16- Noorhosseini-Niyaki, S.A. (2011). Ecological and Biological Effects of Fish Farming In Rice Fields. Regional Congress of Sustainable Management Science-based in Agriculture and Natural Resources. Gorgan University of Agricultural Sciences and Natural Resources, Iran. 21-22 May 2011. pp: 244-250.

17- Noorhosseini-Niyaki, S.A. (2012) Production of Fish Varieties in Rice Fields Simultaneously. National Conference on Food Industries, Young Researchers Club, Islamic Azad University, Quchan Branch, Iran.

18- Noorhosseini-Niyaki, S.A., & Allahyari, M.S. (2010). Problems of Fish Farming in Rice Fields: The Case Study in Guilan Province. The First National Conference of Aquatic Sciences, Islamic Azad University, Booshehr Branch, Iran. 23-24 Feb, 2010. 15p.

19- Noorhosseini-Niyaki, S.A., & Mohammadi, N. (2010). Environmental and Social Benefits of Fish Farming in Rice Fields. The 4th Conference and Exhibition on Environmental Engineering, Tehran University, Iran.

20- Noorhosseini-Niyaki, S.A., & Radjabi, R. (2010). Decline Application of Insecticide and Herbicides in Integrated Rice-Fish Farming: the Case Study in North of Iran. American-Eurasian Journal Agriculture & Environ Science. 8(3): 334-338.

21- Rothuis, A. (1998). Rice-Fish Culture in the Mekong Delta, Vietnam: Constraint Analysis and Adaptive Research. Thesis Submitted for the Award of the Degree of Doctor of Science. Department of Biology, Laboratory of Ecology and Aquaculture, Katholieke Universiteit Leuven, Vietnam. 113p.

22- Saikia. S.K., & Das. D.N. (2008). Rice-Fish Culture and its Potential in Rural Development: A Lesson From Apatani Farmers, Arunachal Pradesh, India Journal Agriculture Rural Development. 6(1&2): 125-131.

23- Salehi, H, Momen-Nia, M. (2006) The Benefits of Fish and Rice Integrated Culture in Iran. Iranian Journal Fish Science. 15(3):97-108.

24- Kathiresan, R.M. (2007). Integration of Elements of a Farming System for Sustainable Weed and Pest Management in the Tropics. Crop Prot 26: 424-429.

10

Assessment of Small Scale Farmers' Skills Regarding Integrated Pest Management (IPM) in District Sargodha-Pakistan

Ejaz Ashraf [1]*, Abu Bakar Muhammad Raza [1], Samiullah [2] and Muhammad Younis [3]

Abstract

A survey study was conducted to assess the knowledge /awareness level in IPM technology among farmers. Four villages were randomly selected from Sargodha district for data collection. Thirteen farmers from each village were selected randomly and sample size was 52 respondents. More than 92% of respondents have no advisory services either from public or private sector. The findings imply that respondents need knowledge for all levels of competence in IPM technology. They need to get high-level of competence for application of this technology in the field. In addition, they have little exposure to long-term training opportunities due to low education level for applications of this technology. More than 77% of respondents think that government agricultural policies and no access to information sources regarding integrated pest management at grass-root level are main constraints. The findings from correlation and regression analyses indicate that age and knowledge/awareness level are negatively correlated. It may be concluded that elder respondents have less adaptability to new ideas and techniques as compared to young respondents. However, training and information, education, and experience play a significant role in enhancing the knowledge/awareness level of respondents in IPM technology.

Keywords:
*Pest Management,
Training, Information,
Knowledge, Awareness*

[1] Assistant Professor, Agriculture. Extension Education. University College of Agriculture, University of Sargodha, Punjab-Pakistan.
[2] Lecturer in Statistics, University College of Agriculture, University of Sargodha, Punjab-Pakistan.
[3] University College of Agriculture, University of Sargodha, Punjab-Pakistan
* Corresponding author's email: ejazashraf60@hotmail.com

INTRODUCTION

Technologies are always being used as essential tool for development and so does Integrated Pest Management (IPM) in modern agriculture. This technology could be used as alternative for curtailing the use of pesticides in crop management. Morse *et al.*, (2000) described that IPM is constantly referred to as one of the main elements in the development of sustainable agriculture. They further described in their study that goals and philosophies of sustainable agriculture and those of IPM have significant similarities. The crisis of uncontrolled use of pesticides required urgent solution, and IPM is regarded as an alternative pest management and solution to the present problems of our growers. The variety of crop protection methods are being used such as crop rotation, biological, and limited use of chemicals.

The IPM technology is driven by many principles, which require careful application of effective pest management. Integrated pest management (IPM) is regarded as the future plant protection model for developing countries and ideal pest management approach (Altieri, 1993). It is expected that IPM has potential to sustain and improve yields, to lower the dependence on pesticides, to decrease the cost of production, and to reduce negative impacts on the environment and human population (Erbaugh *et al.*, 2001). Extensive research is going on in different parts of the world to examine the benefits of this technology, and to check the awareness and knowledge level of farmers for adaptation of IPM technology.

There is a dire need to implement this technology at grass root level for pesticides free crop yields. However, the question is how the goal of implementation could be achieved. The knowledge and awareness about the use of this technology among end users is significant for its application. Moreover, training and information may also play an important role in enhancing the knowledge level of growers in IPM technology.

Knausenberger *et al.*, (2001) pointed out that access to information on IPM practices and implementation can result in IPM becoming a sustainable crop protection method in Africa.

Kyamanywa (2001) noted that in Uganda, extension agents recommend the use of chemicals and overlook cultural tactics due to limited information on its application. The role of extension workers is more important than any other factor in adoption of IPM technology at grass root level. Providing knowledge and information to growers are the core responsibilities of extension workers. The extension system could play a significant role in proper adoption of IPM approaches and for sustainable agricultural. Norris *et al.*, (2003) indicate that the economic reliability of IPM practices is particularly crucial, and to a great extent, determines adoption. IPM practices must reduce the risk of crop loss or encounter rejection by farmers.

The modern research tells that pest management is crucial for any farming system. Plants and animals must be protected from damage caused by insects, weeds, nematodes and other pathogens. Due to relaxed policies in pesticide use regulations in developing countries, farmers are continuously relying on chemical without considering their negative effects on human and animal health and also for environment. It is important to create awareness and boost up knowledge level of growers for using IPM technology as a solution to the prevailing problems in agriculture sector. IPM is regarded as a practical alternative in pest management for overcoming issues that arise from increased pesticide use (Afreh-Nuamah, 2001).

MATERIALS AND METHODS

A survey study was conducted to assess the knowledge /awareness level in IPM technology of small scale farmers. Erbaugh *et al.*, (2001) conducted a similar study to evaluate farmers' knowledge and awareness regarding IPM in a collaborative research support project in Uganda. The present study was conducted in Sargodha district of Punjab province of Pakistan. The two stage random sampling procedure was used to collect data from respondents. Four villages were randomly selected from Sargodha district and then13 farmers from each village were selected randomly. Therefore, the sample size was 52 respondents. A survey questionnaire was used as an instrument for this study. The instrument was pre-tested on 10 farmers and then finalized by a panel of experts after making number of corrections. In addition, instrument was also tested for reliability by calculating the Cronbach's alpha using the data obtained from pilot study and found reliable. The instrument consisted of four sections demographic profiles of respondents, constraints in using IPM, training and information, and knowledge/awareness level of respondents.

Objectives of the Study

The study was conducted to achieve the following objectives.

1- To describe the demographic profiles of the respondents

2- To identify the knowledge/awareness level of respondents in specific concepts of IPM technology.

3- To identify the opportunities of training and information in IPM technology for respondents.

4- To describe constraints those limit the applications of IPM technology.

5- To identify the relationship among the variables such as training/ information for IPM technology, demographic profiles, and knowledge/awareness level of respondents in specific concepts of IPM technology.

Data Analysis

The collected data was coded and entered into computer for further analysis. The Statistical Package for Social Sciences (SPSS) 15.0 was used for analysis. The descriptive statistics such as mean, standard deviation, frequency distribution and percentages were computed for general description of the data. The multiple regression analysis was also used to predict how much of variance in dependent variable of knowledge / awareness level accounted for by independent variables of training/information and demographic profiles of the respondents. Multiple regression procedures also make it possible to observe the relationships between each of independent variable and dependent variable while controlling for other variables in the model (Urdan, 2001). Inter-correlations also computed to show strength of mutual relationships between the variables.

RESULTS, DISCUSSIONS, AND CONLUSIONS

The demographic profiles were age, education, years of experience in farming, and advisory services. Average age of the respondents was

Table 1: Means, standard deviations, and ranks of knowledge/awareness level in specific Concepts of IPM technology as perceived by the respondents

Knowledge / Awareness level in IPM Technology	N	Mean*	SD	Rank
Harmful effects of Pesticides use	52	3.58	0.99	1.0
Prevention is better than cure	52	3.27	1.43	2.0
Important Insect/Pests Knowledge	52	2.77	0.73	3.5
Pruning and Thinning can minimize pest population	52	2.77	1.00	3.5
Resistant varieties against different Insect/Pests	52	2.56	0.61	5.0
Timely sowing can reduce Insect/Pests population	52	2.52	0.80	6.0
Damage Symptoms of Insect/Pests	52	2.48	0.80	7.0
Insect/Pest/Control	52	2.38	0.77	8.0
Susceptible Stages for Insect/Pests attack on plant	52	2.17	0.90	9.0
Use of light traps Or Pheromone traps	52	2.12	1.04	10.0
Inter-cropping and crop rotation is essential for pest control	52	2.04	0.93	11.0
Techniques of Control Insect/Pests	52	1.96	0.79	12.0
Other methods to control attack of Insect/Pests	52	1.75	0.84	13.0
Pesticides as a last option to control attack of Insect/Pests	52	1.67	0.78	14.0
Alternative use of pest control measures are effective	52	1.63	0.82	15.0
Use of Fertilizers to control Insect/Pests population	52	1.62	0.57	16.5
Options available to control Insect/Pests other than pesticides	52	1.62	0.72	16.5
Selective use of Pesticides to control attack	52	1.58	0.80	18.0
Inter-cultural Practices to control attack of Insect/Pests	52	1.37	0.56	19.0

*Mean: 1= None, 2=Low, 3=Moderate, 4=High, 5=Very high

46 years. More than 67% of the respondents were of the age from 40 to 59 years. Of the 52 respondents, almost 35% were educated between grade eight (Middle) to grade ten (Metric) and only 10% were above metric. These results indicate that respondents belong to mature age group with low education level. Approximately 46% of respondents have farming experience from 21 to 30 years. Almost 19% of respondents have 31 to 40 years farming experience Unfortunately more than 92% of respondents have no advisory services either from public or private sector.

The results indicated that respondents have moderate knowledge/awareness level in two concepts of IPM technology such as "Harmful effects of Pesticides use (3.58) and Prevention is better than cure (3.27)". The respondents have low level of knowledge and awareness in "Important Insect/Pests (2.77) and Pruning and Thinning can minimize pest population with mean of (2.77)". Of the 19 specific concepts of IPM technology, the respondents were considered having "No" knowledge/awareness in 8 specific concepts of IPM technology as shown in the following table.

These findings imply that respondents have none to moderate knowledge and awareness in all specific concepts of for IPM technology which does not provide enough competence for applications of this technology in the field. The respondents need to get high-level of competence for applications of this technology. Ashraf (2007) reported similar findings in his doctoral study

while assessing in-service educational needs of agricultural officers for adaptation of remote sensing technology for precision agriculture in the province of Balochistan-Pakistan.

The respondents were asked to rate the accessibility of training and information opportunities on a 5-point Likert scale ranging from none to very high (1-5). The means ranged from 1.21 (none) for "ETL/EIL knowledge" to 3.13 (moderate) for "other farmers as source of information" shown in the table given below.

The respondents described that they have "none" to "moderate" access to the training/information opportunities for IPM and its applications in agriculture. These findings indicate that respondents have limited access to training/information opportunities. The results further confirm that extension system in Pakistan does not provide enough training and information to clients for applications of IPM technology in the field. One of the reasons is, the extension officers themselves have limited options of training for this technology. It also implies that respondents have limited interaction with other professionals/researchers for applications of IPM technology. Consequently, they have little exposure to the long-term training opportunities due to low educational level.

The respondents also pointed out the constraints in applications of IPM technology. Few of them feel that inputs and finance are major constraints However, 77% respondents think that government agricultural policies and no access to information sources regarding integrated pest management

Table 2: Means, standard deviations, and ranks for training and information opportunities for IPM technology

Sources of Training/Information for IPM	N	Mean*	SD	Rank
Other farmers	52	3.13	0.79	1
Electronic Media	52	2.50	0.64	2
Print Media	52	1.88	0.78	3
Interactions with researchers	52	1.65	0.81	4
Extension worker	52	1.50	0.61	5
Training Sessions	52	1.48	0.77	6
Bulletins	52	1.46	0.58	7
Seminars	52	1.44	0.70	8
Publications	52	1.36	0.60	9
Short Courses	52	1.28	0.54	10
ETL/EIL Knowledge	52	1.21	0.41	11

*Mean: 1=None, 2=Low, 3=Moderate, 4=High, 5=Very high

Table 3: Regression analysis for Knowledge/awareness (DV)

Variable	Coefficients	S.E.	t	Sig.
Training and Information	1.044	0.190	5.482*	<0.001
Experience	0.804	0.142	5.663*	<0.001
Education	3.680	0.795	4.628*	<0.001
Age	-0.614	0.162	-3.779*	<0.001
Constant	18.570	6.025	3.081*	0.003

R^2 =.783; Adj. R^2 =.765; df= 4, 47; F= 42.505 *Significant at .05 level

at grass-root level are main constraints.

A regression model was used with independent variables of training/information, experience, education, and age of respondents. The model was statistically significant p < 0.0001 and explained 78.34% of the variance for knowledge/ awareness level. The results indicated that independent variables such as training and information, experience, education, and age played a significant role in knowledge /awareness level among respondents in specific concepts of IPM technology. In addition all demographic variables were statistically significant.

The findings from correlation and regression analyses also indicate that age and knowledge/ awareness level are negatively correlated. It may be concluded that elder respondents have less adaptability to new ideas and techniques as compared to young respondents. However, training and information, experience, education, and age play major role in enhancing knowledge/ awareness level of respondents in IPM technology. Hussain *et al.,* (2011) concluded similar findings in their study that in Pakistan education plays an effective role in the adoption of IPM technology. They further recommended that government may take actions to upgrade the education as well as training programs in IPM for cotton producers.

RECOMMENDATIONS

The following are few recommendations for implementation and dissemination of this technology at grass-root level in Pakistan.

1- Continuous training programs are required for propagation of this technology among small scale farmers.

2- Refresher courses are the keys for success

3- Doing IPM practices onsite on farmers land.

4- Permanent coordination among farmers and researchers is required for the solution of the problems.

5- Educational institutions of higher learning need to include courses and offer training programs for Extension staff for up-gradation of their skills in IPM technology.

6- The Government should be responsible for implementing agricultural policies on priority bases.

REFERENCES

1- Afreh-Nuamah, K. (2001). The status of IPM in Ghana. Retrieved December 14, 2010 from http://www.ag.vt.edu/ail/proceedings/ghana.htm

2- Altieri, M.A. (1993). Crop protection strategies for subsistence farmers, Boulder, Colorado: Westview Press.

3- Ashraf, E. (2007). In-service educational needs of agricultural officers for adaptation of remote sensing technology for precision agriculture in the province of Balochistan, Pakistan, Unpublished doctoral dissertation, Mississippi State University at Mississippi State.

4- Erbaugh, J.M. (2001). Activities of the IPM CRSP in Sub-Saharan Africa. Retrieved December 14, 2010, from http://www.ag.vt.edu/ail/proceedings/usaid.htm

5- Hussain, M., Zia, S., & Saboor, A. (2011). The adoption of integrated pest management (IPM) technologies by cotton growers in the Punjab. Journal of Soil Environ. 30(1): 74-77.

6- Knausenberger, W.I., Schefers, G.A., Cochrane, J., & Gebrekidan, B.(2001). Africa IPM-Link: An initiative to facilitate IPM information networking in Africa. Retrieved December 14, 2010 from http://www.ag.vt.edu/ail/proceedings/usaid.htm

7- Kyamanywa, S. (2001). Current status of integrated pest management in Uganda. Retrieved December 14, 2010 from http: //www.ag.vt.edu/ail/proceedings/usaid.htm

8- Morse, S., McNamara, N., Acholo, M. & Okwoli, B. (2000). Visions of sustainability: stakeholders, change and indicators. Hampshire, England: Publishing Company.

9- Norris, R.F., Caswell-Chen, E.P.; Kogan, M. (2003). Concepts in integerated pest management. New Jersey: Pearson Education, Inc.

10- Urdan, T. C. (2001) Statistics in plain English. New Jersey: Lawrence Erlbaum

Association, Publishers.

11

Investigating Consumers' Willingness to Pay for Organic Green Chicken in Iran (Case Study: Rasht City)

Mohammad Kavoosi Kalashami [1*], Morteza Heydari [2] and Houman Kazerani [2]

Abstract

Health and safety are important factors in today's life. Most of studies show that consuming the chicken that has anti biotic caused different diseases like digestive vessel cancer. Attention to controlled use of antibiotics can play a key role in society health which considered in the production of green chicken. Planning for increasing the production of mentioned chicken needs the investigation of consumers' WTP. So, using double bound contingent valuation method and logit model, present study estimates consumers' WTP for green chicken in Rasht city. Among the explanatory variables applied in logit model, income and education level had positive and significant effects on WTP for green chicken. Results revealed that average WTP for a kg of green chicken equals to 37279 Rials, and because this WTP does not compensate the production costs, government protection such as Green subsidy should be considered in order to expand green products consumption in Iran.

Keywords:
Willingness to Pay, Organic Products, Green Chicken, Consumers' Preference, Rasht City

[1] *Lecturer, Islamic Azad University, Rasht Branch, Rasht, Iran.*
[2] *MSc students of Agricultural Management, Islamic Azad University, Rasht Branch, Rasht, Iran.*
* *Corresponding author's email: mkavoosi@ut.ac.ir*

INTRODUCTION

By improving science and new technologies like changing the growth speed of poultry, especially chicken by means of genetic engineering, changing the food consumption pattern and creating a suitable farm environmental situation, in last three decades, caused a big revolution in chicken production. Production increase solved the food shortage problem in many of developed countries and developing countries, but new problems like new deceases and reduction of food quality are created. Diseases increase like different kinds of cancers which is derived from consuming the agricultural products caused worldwide view change toward the use of more organic food.

In reality, unfavorable advantages and remained effects of consuming different kind of chemicals, hormones and antibiotics in nutrition production in different developed countries, caused agriculture develops in opposite side of modern method and technology, in which there is a prohibition in consuming chemical materials or artificial things in production process named organic production.

Antibiotics are groups of chemical compounds that produced biochemically by plants or micro organism like spores and have antibacterial characteristics and prevent bacterial growth and its less doze (5ppm) in a period can increase the poultry's growth (Jabarzadeh, 2011). Consuming antibiotics causes two main problems in human health that consists of remained antibiotic in bestial products and pathogen's resistance against antibiotic. Remained antibiotics in bestial products caused allergic response, fever, squirt, charley horse of gastric muscle, harmful effects on metabolism in digestion system and in long run consumption caused cancer. According to the previous studies, different kinds of cancer have a direct and close relation with consuming protein and antibiotics. In Iran about 200,000 cases of cancers recorded annually that about 36 percent of these cancers related digestion system and liver cancer (Akbari, 2000).

In Iran more than 1000 million tones, chicken meat produces annually (FAO, 2011). In recent years nearly about 120 production unit of green

chicken activated among these units, 12 centers are producing green chicken in Guilan province (Ministry of Jihad-Agriculture, 2011). Omitting antibiotics in chicken production raise costs and normally this green product should be sold with higher price in markets. In this case investigating consumers' preference and willingness to pay (WTP) for green chicken helps producers to price their commodity better and analyze the possibility of production increase.

WTP can be defined as the money that paid to firm and is the different between surplus pay of consumers before and after the improvement in one characteristic of nutrition products. (Rodriguez and et al., 2007).

Contingent Valuation Method (CVM) had been used increasingly for valuing green products throughout the world. Mafi and Saleh (2009) estimated the consumers' WTP for organic vegetable and cucumber in Guilan and Tehran provinces of Iran. Results revealed that income and knowledge about cancers had positive and significant effect on WTP. Tagbata and Sirieix (2008) studied the effect of organic label on consumer's WTP in France. Results showed that half of the Consumers are sensitive to organic label in food buying. Rodriguez and et al., (2007) evaluated consumers' WTP for organic products in Argentina. Their findings showed that consumers would pay more money for buying organic products something between 6 to 200 percentages.

Considering the importance of the cognition of consumers' WTP for green chicken in production increase plans and marketing strategy, present study investigated the WTP for green chicken in Rasht city using CVM and survey analysis.

MATERIALS AND METHODS

In order to estimates the consumers' WTP for green chicken applying CVM, 200 random individuals in Rasht city selected and requested explanatory variables obtained through questionnaires filled by sample individuals. Suppose the utility function of Rasht city consumers as below:

$$U(Y,S) \tag{1}$$

In which, U is indirect utility function, Y is individual's income and S is the vector of individual's social-economic characteristics. Each consumer agrees to pay some of his or her income for buying green chicken that we call this amount A, only if consuming green chicken increases his or her utility more than the situation in which this consumption does not occurred. Below equation express mentioned situation:

$$U(1,Y-A;S)+\varepsilon_1 \geq U(0,Y;S)+\varepsilon_0 \tag{2}$$

In above relation, ε_1 and ε_0 are random variables with average of zero that distributed accidently and independent of each other (Khodaverdizadeh and *et al.,* 2011). The increase in individual's utility because of green chicken consumption defined as below (Akbari and *et al.,* 2007):

$$\Delta U = U(1, Y - A; S) - U(0, Y; S) + (\varepsilon_1 - \varepsilon_0) \tag{3}$$

According to the per-test statistical distribution, double bound approach has been used in CVM questionnaires designing. CVM questions in double bound approach followed a binary answer (accept or deny the bid) by individual (Lee and Han, 2002). For estimating valuation function in CVM, Logit functional form used widely (Amirnejhad *et al.,* 2006). In valuation model, Logit functional form used to study the different explanatory variables effects on WTP of individuals for consuming green chicken.

In Logit model the probability of accepting the bids by individuals defined as below:

$$p_i = F_\eta(\Delta U) = \frac{1}{1 + \exp(-\Delta U)} = \frac{1}{1 + \exp\{-(\alpha - \beta A + \gamma Y + \theta S)\}} \tag{4}$$

In which, $F\eta(\Delta U)$ is aggregate distribution function with standard logistic difference and in which explanatory variables like income, bid, age, gender, family size and education level had been used for estimating valuation function. Also, β, γ and θ are regression coefficients which are expected to be β≤0, γ > 0 and θ > 0, respectively. Logit model could be estimated in linear or logarithmic form. The interpretation of two parameter is important in logit model results include elasticity and marginal effect. The elasticity of k explanatory variable (X_k) is as below (Hayati and *et al.,* 2011):

$$E = \frac{\partial(B'X_K)}{\partial X_K}\frac{X_K}{B'X_K} = \frac{e^{B'X}}{(1+e^{B'X})^2}B_K\frac{X_K}{(B'X_K)} \tag{5}$$

Elasticity of an explanatory variable explained the percentage change in the probability of bid acceptance for green chicken buying by individual when X_k amount changed by one percentage. Also, the marginal effect showed the percentage change in the probability of bid acceptance for green chicken buying by individual when X_k amount changed by one unit.

$$ME = \frac{\partial Pi}{\partial x_k} = \frac{\exp(B'X)}{(1 + \exp(B'X))^2}.B_k \tag{6}$$

In above relation, the extent of change in the probability of bid acceptance depends on the initial probability and initial value of independent variables and their coefficients. When the explanatory variable is a dummy variable, the following relation is used for computing the marginal effect (Khodaverdizadeh and *et al.,* 2011):

$$ME = p(y = 1|x_k = 1, X^*) - P(Y = 1|x_k = 0, X^*) \tag{7}$$

Marginal effect for a dummy variable equals to change in the probability of bid acceptance (Yi = 1) as a result of changing dummy variable amount (X_k) from 1 to 0, while other variables amounts held constant at (X*) levels.

For calculating the maximum WTP of sample individuals, considering the linear Logit functional form, below equations had been used (Haneman, 1984):

$$u(1, y - A; s) = u(0, y; s)$$
$$v(1, y - A; s) + \varepsilon_1 = v(0, y; s) + \varepsilon_0 \Rightarrow \Delta v = 0 \tag{8}$$

We can write indirect utility difference as below:

$$v(h, y - A; s) = \alpha_h + \beta y + \varepsilon_h \quad , \beta \succ 0, \ h = 0,1$$
$$v(1, y - A; s) = \alpha_1 + \beta(y - A) + \varepsilon_1$$
$$v(0, y; s) = \alpha_0 + \beta y + \varepsilon_0$$
$$\Delta v = v(0, y; s) + \varepsilon_0 - v(1, y - A; s) - \varepsilon_1 = (\alpha_0 - \alpha_1) + \beta A + \eta$$
$$\tag{9}$$

Considering that the average amount of equals 0, above equation could be rewrite as below:

Table 1. Descriptive statistics of sample's characteristics.

Characteristic	Average	Maximum	Minimum	SD	Mode
Age	42.36	77	20	13.23	36
Education level*	4.28	7	2	1.24	5
The number of family	3.74	8	1	1.14	4
chicken consumption**	3.63	12	1	1.64	3

*Education level categorized to 7 levels.
**The chicken consumption during a week

$$\Delta v = v(0, y; s) - v(1, y - A; s) = (\alpha_0 - \alpha_1) + \beta A$$

(10)

From which, maximum amount of individual's WTP for a kg of green chicken equals $-\frac{\alpha_0 - \alpha_1}{\beta}$. Requested data set obtain from a survey that was done in early 2011 in the city of Rasht.

RESULTS AND DISCUSSION

Investigated sample include 200 citizen of Rasht city that was selected randomly. Out of total sample individuals, there were 29.5 percent (59) women and 70.5 percent (141) men. Important descriptive statistic of investigated sample reported in table 1.

Out of total sample size, 37.5 percent knew the advantages of green chicken consumption, while 62.5 percent didn't know anything about the advantages of green chicken. To evaluate the consumers' views about the advantages of green nutrition, four questions based on likert scale asked from each individual. Sample's answers to these questions reported in table 2.

Using scoring approach, individuals' answers to attitude questions applied for constructing individual attitude variable toward green chicken consumption advantages and this variable used in regression analysis of determining effective variables on acceptance of bids, proposed for a kg of green chicken. Three bids used in double bound approach of CVM which are include 37000, 44000 and 52000 Rials per a kg of green chicken. These bids proposed based on full cost and two scenarios of considering 20 and 40 percent margins. For estimating individuals' WTP for a kg of green chicken, explanatory variables of proposed price (BID), education level (EDU), family size (FN), dummy variable of consumption experience (CE), chicken consumption meals per week (CN) and attitude variable of green chicken consumption advantages (SAS) had been considered. In order to estimate Logit binary model, at first multicollinearity existence among explanatory variables had been investigated. (Table 3)

According to this fact that there isn't any pair of numbers greater than 0.5 in each row of above table, it could be concluded that there is no multicollinearity among the explanatory variables of this study.

Table 2. The frequency of sample's answer to four attitude questions.

Attitude question	A	B	C	D	E
The development of green nutrients production firms should have priority in the development programs of Iran.	30	71	96	2	1
Although the price of green product is higher than the same common product but green products should have a greater share in my family's consumption basket.	39	88	69	4	0
In my opinion the value and utility of green nutrients and common nutrients are equal.	2	14	61	96	27
Although green nutrients consumption helps us to improve our health but I do not want to pay more for buying these products.	8	51	80	45	16

A: completely agree B: agree C: indifferent D: disagree E: completely disagree

Table 3. Principle component test results.

Explanatory variable	BID	EDU	FN	CE	CN	SAS
1	0.84	0	0	0	0	0
2	0.14	0.01	0.003	0.0001	0	0.71
3	0.01	0.0001	0.008	0	0.95	0.0006
4	0.0006	0.25	0.65	0.001	0.02	0.19
5	0.003	0.59	0.34	0.002	0.02	0.08
6	0.002	0.14	0.0008	0.99	0.01	0.01

Table 4. The results of Logit model estimation.

Variables	Coefficients	SD	t-statistics	Elasticity
BID	-0.31×10^{-2}	0.39×10^{-3}	-7.86*	-5.87
EDU	0.45	0.13	3.3*	0.95
FN	-0.19	0.12	1.55	-0.33
CE	0.49	0.31	1.56	0.084
CN	0.25	0.083	3*	0.44
SAS	0.47	0.074	6.41*	3.2
Constant	2.68	1.43	1.87**	-

* Significant at 1%
** Significant at 5%

For evaluating the effects of explanatory variables on binary dependent variable (accepting bid or denying) Logit model estimated using maximum likelihood estimator. The results of model estimation reported in table 4.

The signs of two explanatory variables include BID and FN was negative that showed the reverse effect of mentioned variables on the acceptance of proposed price for a kg of green chicken. The t-statistics of BID variable showed that its negative effect on dependent variable was significant at 1 percent. The elasticity of BID showed that a percent increase in BID amount decrease 5.87 percent the probability of accepting the proposed price by consumer. Explanatory variables include EDU, CE, CN and SAS had positive signs and had direct effects

on proposed price acceptance by individuals. According to the t-statistics, direct and positive effects of EDU, CN and SAS were significant on 1 percent. The elasticity amounts of these variables showed by 10 percent increase in EDU, CN and SAS levels, the probability of proposed price acceptance for a kg of green chicken increase 9.5, 4.4 and 32 percent, respectively. Also, the marginal effect calculation for each variable reported in table 5.

The marginal effect of BID showed that 10 thousand Rials increase in proposed price amount reduces the probability of WTP and bid acceptance by 0.51 units. The negative marginal effect of FN revealed that one unit increase in its level reduces the probability of bid acceptance by 0.03 units. Positive marginal effect of EDU,

Table 5. Marginal effect of different explanatory variables.

Variable	Type of variable	Case Value	Marginal Effect
BID	Continuous	4207.5	-0.51×10^{-3}
EDU	Ranked	4.28	0.07
FN	Continuous	3.74	-0.03
CE	Dummy	0	0.08
CN	Continuous	3.63	0.04
SAS	Continuous	14.75	0.07

* Significant at 1%
** Significant at 5%

CE, CN and SAS showed that increase in their amounts would increase the probability of proposed price acceptance by individuals.

The LR (Likelihood Ratio) calculated statistic for Logit model equals 145.29 with 0 probability value which showed the significance of estimated Logit model. Logit Model Percentage of Right Prediction equals 71 percent which insist on high prediction power of model.

Applying Haneman (1984) method for calculating WTP showed that the amount of individual's average WTP for a kg of green chicken equals 37279 Rials. This willingness to pay amount is close to the full cost amount for a kg of green chicken and suggests that producers could not gain high margins by pricing more in markets because less consumers would buy this product in higher prices.

CONCLUSION

According to the increase of cancer diseases in Iran and the importance of society health, expanding the production and consumption of organic products is inevitable. Consumers' knowledge and information increase about the advantages of green products and the effects of antibiotics on health would change their tendency toward these products.

Advertising green products is in different ways, using public sector marketing system besides private marketing would change consumers' tastes and attitudes toward these products. Developing the standards for green products, intensifying the control process in food production firms and allocating subsidy to green products by government would increase these products share in consumers' consumption baskets.

REFERENCES

1-Ahmadvand, M.R. & Najafpor, Z.A. (2003). Calculation and analysis of physical indicators of comparative advantage in production of oilseeds in Iran, Iranian Journal of Research and Economic Policies, No. 37 and 38.

2- Akbari, M.A. (2000). Cancer Research Shahid Beheshti University, the Third Congress of breast cancer, Tehran, Iran.

3- Akbari, N.A., Khoshakhlagh, R., Sameti, M. & Shahidi, A. (2007). Estimating consumers' willingness to pay for quality increased bread in Esfahan city, Iranian Journal of Agricultural Economics, 3(1): 89-113

4- Amirnejhad, H. Khaliliyan, S. & Osarehe, M.H. (2006). Determining the preservation and recreational values of the Noshahr Sisangan Forest Park using willing to pay, Iranian Journal of Construction and Research. 72:15-24

5- Becker, T. (2006). Consumer's Attitude and Behavior towards Organic Food cross-cultural study of Turkey and Germany, Master Thesis, Institute for Agricultural Policy and Markets, Stuttgart,Hohenheim.

6.- FAO (2011). Available at www.fao.stat.org

7- Hayati, B.A., Ehsani, M., Ghahremanzadeh, M., Raheli, H. & Taghizadeh, M. (2010). Factors affecting the willingness to pay for Elgoli Tabriz park visitors, Iranian Journal of Economics and Agricultural Development (Agricultural Science and Technology). 24(1): 91-98.

8- Haneman, W.M. (1984). Welfare evaluation in contingent valuation experiments with discrete responses. American Journal of Agricultural Economics, 71(3): 332- 341.

9- Jabarzadeh, M. (2011). The use of antibiotics in livestock and poultry feed and its effect on human health, livestock magazine, No.140, page 34.

10- Khodaverdizad, M., Kavoosi Kalashami, M., Shahbazi, H., & Malekian, A. (2011). Estimating the ecotourism value of Mahabad Saholan cave using contingent valuation, Geography and development, 23: 203-216.

11- Lee, C. & Han, S. (2002). Estimating the use and preservation values of national parks tourism resources using a contingent valuation method, Tourism Management 23: 531-540.

12- Ministry of Jihad-Agriculture (2011). Poultry industry statistics.

13- Mafi, H. & Saleh, A. (2009). Estimating the willingness to pay for organic vegetables and cucumber in Guilan and Tehran provinces of Iran, Sixth Iranian Conference of Agricultural Economics, Mashhad, Iran.

14- Molai,M. ,Sharzaei,G. & Yazdani, S.(2009). Extraction the information from the questionnaires effects on willingness to pay results in contingent valuation, Journal of Economic Research ,90:159-181.

15- Nyung, H.J. & Hwan, S.M. (2004). Measuring Consumers' Value for Organic-beef using contingent valuation method, Journal of rural, 27: 95-110.

16- Rodriguez, E., Lacaze, V. & Lupin, B. (2007).

Willingness to pay for organic food in Argentina, Evidence from a consumer survey, Contributed Paper prepared for the presentation at the 105th EAAE Seminar 'International Marketing and International Trade of Quality Food Products' Bologna, Italy.

17- Tagbata, D. & Sirieix, L. (2008). Measuring consumer's willingness to pay for organic and Fair Trade products, International Journal of Consumer Studies, 32: 479–490

12

Price Transmission Analysis in Iran Chicken Market

Seyed Safdar Hosseini [1], Afsaneh Nikoukar [2] and Arash Dourandish [3]

Abstract

Over the past three decades vertical price transmission analysis has been the subject of considerable attention in applied agricultural economics. It has been argued that the existence of asymmetric price transmission generates rents for marketing and processing agents. Retail prices allegedly move faster upwards than downwards in response to farm level price movements. This is an important issue for many agricultural markets, including the Iranian chicken market. Chicken is an important source of nutrition in Iranian society and many rural households depend on this commodity market as a source of income. The purpose of this paper is to analyze the extent, if any, of asymmetric price transmission in Iran chicken market using the Houck, Error Correction and Threshold models. The analysis is based on weekly chicken price data at farm and retail levels over the period October 2002 to March 2006. The results of tests on all three models show that price transmission in Iranian chicken market is long-run symmetric, but short-run asymmetric. Increases in the farm price transmit immediately to the retail level, while decreases in farm price transmit relatively more slowly to the retail level. We conjecture the asymmetric price transmission in this market is the result of high inflation rates that lead the consumers to expect continual price increases and a different adjustment costs in the upwards direction compared to the downwards direction for the marketing agents and a non-competitive slaughtering industry and that looking for ways to make this sector of the chicken supply chain more competitive will foster greater price transmission symmetry and lead to welfare gains for both consumers and agricultural producers.

Keywords:
Asymmetric Price Transmission, Iran Chicken Market, Houck Model, Error Correction Model, Threshold Model

[1] *Professor, Department of Agricultural Economics, College of Agriculture, University of Tehran,Karaj,Iran.*

[2] *Assistance professor, Payam-e-noor University, Khorasan Razavi, Mashhad, Iran.*

[3] *Assistance professor, Department of agricultural economics, College of Agriculture, Ferdowsi University of Mashhad, Iran.*

* Corresponding author's email: hosseini_safdar@yahoo.com

INTRODUCTION

Over the past three decades, vertical price transmission analysis has been the subject of considerable attention in applied agricultural economics (Meyer and Von Cramon-Taubadel, 2004 and Goodwin, 2005). It has been argued that the existence of asymmetric price transmission may generate rents for marketing and processing agents as retail prices move faster upwards than downwards in response to farm level price movements. This paper details our study of the price transmission process for the Iranian chicken where this issue has important economic welfare and political implications.

Chicken meat is an important commodity in Iran's economy. Chicken provides 50 percent of per capita meat consumption and is appreciated by Iranian consumers as a cheap source of protein in comparison with beef and lamb. Chicken is used by all income classes and its consumption has grown more than 250 percent since the 1979 revolution. Per capita chicken consumption in Iran is currently around 16.9 kg/year.

The size of the Iranian chicken market is about 1.2 million metric tons. There are more than 15,000 active producers producing 900 million live birds per year for this market. 89 percent of the farms are private farms, 8 percent of them belong to the cooperatives and 3 percent of them belong to the government. There are also 177 chicken slaughterhouses with a total annual slaughter capacity of 912 million live chickens. 76 percent of slaughterhouses are private firms, 10 percent of them belong to the government and 14 percent of them belong to the cooperatives. 67 percent of chicken production takes place in 10 provinces of Iran. In the other hand, 100 slaughter-houses with 66 percent of slaughtering capacity are located in 8 provinces; Tehran, Isfahan, Khorasan, Fars, Eastern Azarbaijan, Western Azarbaijan, Ardebil and Yazd. However, only 47 percent of the capacity of chicken production is in these provinces. Thus not only the number of slaughter-houses is less than the number of farms but also there is not a balance between production capacity and process capacity in different regions of the country. Because of this imbalance, some producers transport their live chicken for slaughtering to the regions that have excess capacity, for example from Mazandaran, Zanjan, Ghazvin and Qom to Tehran. On the other hand, wholesalers and middlemen transport chicken meat to the regions those have excess demand. In Iranian chicken market there are 515 wholesalers and 49000 retailers. There are not accurate statistics about the ownership of the wholesale and retail firms but our field operations in Tehran province showed that almost all of the wholesale firms belong to the owners of the slaughter-houses.

In May 2003, the Iranian Government introduced a price stabilization scheme for chicken. Under this scheme a ceiling and floor price are determined administratively every 2 to 4 months based on cost of production (including an 8% profit margin for producers). When market price falls below the floor, the government pays for the live purchase, slaughter, freezing and storage of chicken in an attempt to lift market price to at least the floor level. The buffer stock operations are carried out by a public-private organization which stores the chicken in one of its many of storage facilities maintained throughout the country. When the price moves above the ceiling, the public-private organization releases frozen chicken onto the market in an attempt to bring the market price down to at least the ceiling level. This policy is potentially a significant factor affecting price transmission and hence needs to be considered in any analysis of the asymmetry of price transmission.

The principal objective of this paper is to estimate farm-to-retail price transmission elasticities in the Iranian chicken market and to explore the existence of asymmetric price transmission. A number of alternative methods have been proposed for analyzing the existence of asymmetric price transmission. And, according to Meyer and Von Cramon-Taubadel (2004), different methods may lead to different conclusions. Thus we propose to explore three alternative methods, including the Houck approach, the Error Correction model and the Threshold model. In our analysis we employ weekly data over the period October 2002 to March 2006 pertaining to farm and retail prices of chicken from Iran.

MATERIALS AND METHODS

In this Section, three alternative models for analyzing asymmetric price transmission are discussed.

1. Houck Model

Wolffram (1971) was the first to propose a variable-splitting technique in the first differences of prices to estimate the asymmetry in price transmission. Houck (1977) proposed a modification to exclude the initial observations because, when considering differential effects, the level of the first observation will have no independent explanatory power. Further modifications to this approach were introduced by Ward (1982) to include lagged exogenous variables and by Boyd and Brorsen (1988) who also used lags to differentiate between the magnitude and the speed of transmission. The modified Houck Approach has been widely used (e.g. Kinnucan and Forker, 1987; Bailey and Brorsen, 1989; Hahn, 1990; Mohanty et al., 1995; Aguiar and Santana, 2002; Capps and Sherwell, 2005). Equation (1) shows the modified Houck model for the Iranian chicken market:

$$\Delta RP_t = \alpha_0 + \sum_{i=0}^{L1} \alpha_{1,i}\Delta FP_{t-i}^- + \sum_{i=0}^{L2} \alpha_{2,i}\Delta FP_{t-i}^- + \alpha_3 D_{2003} + \varepsilon_t$$

(1)

where:

$\Delta RP_t = RP_t - RP_{t-1}$ is the observation-to-observation difference of chicken meat price at the retail level;

ΔFP^+_{t-1} and ΔFP^-_{t-1} are the increases and decreases of the live chicken price at the farm respectively;

D_{2003} is a dummy variable for the Government's market adjustment policy. It equals 0 for observations prior to the introduction of the policy on May 9, 2003 and 1 thereafter.

α_0, $\alpha_{1,i}$, $\alpha_{2,i}$ and α_3 are the coefficients to be estimated. The $\alpha_{1,i}$ coefficients represent the impact of farm price increases on retail price and the $\alpha_{2,i}$ coefficients represent the impact of farm price decreases on retail price;

L_1 and L_2 are the maximum lag lengths for farm price increases and decreases respectively; and

ε_t is the random error term.

After estimating equation (1), two tests may be performed for the existence of price transmission asymmetry. They are tests with respect to the magnitude and speed of price transmission. The magnitude test for asymmetric price transmission can be represented by the null hypothesis H_0 in equation (2) below.

$$H_0: \quad \sum_{i=0}^{L1} \alpha_{1,i} = \sum_{i=0}^{L2} \alpha_{2,i} \quad (2)$$

A rejection of H_0 is evidence for asymmetry in the magnitude of price transmission.

The speed test for asymmetric price transmission can be represented by the null hypothesis H_0 in equation (3):

$$H_0: \alpha_{1,1} = \alpha_{2,1}, \alpha_{1,2} = \alpha_{2,2}, \ldots, \alpha_{1,L1} = \alpha_{2,L2} \quad (3)$$

A rejection of H_0 is evidence for asymmetry in the speed of price transmission.

2. Error-Correction Model

Von Cramon-Taubadel and Loy (1996) proposed testing for asymmetric price transmission between co-integrated price series by using an Error Correction Model (ECM) extended by incorporating asymmetric adjustment terms. Scholnick (1996), Bornstein et al., (1997) and Capps and Sherwell (2005) have each used this approach to test asymmetric price transmission.

To use this approach, we first estimate the co-integration relationship represented in equation (4):

$$RP_t = \lambda_0 + \lambda_1 FP_t + \lambda_2 D_{2003} + e_{RF,T} \quad (4)$$

Here RP_t is the retail price of chicken meat, RP_t is the farm price of live chicken and D_{2003} is a dummy variable for the Government's market adjustment policy (equals 1 after the introduction of the policy on May 9, 2003 and 0 otherwise). After estimating (4), the lagged co-integrating residuals $e_{RF,T-1}$ are split into positive and negative phases used in estimating the ECM for the Iranian chicken market:

$$\Delta RP_t = \alpha_0 + \sum_{i=0}^{L1} \alpha_{1,i}\Delta FP_{t-i}^+ + \sum_{i=0}^{L2} \alpha_{2,i}\Delta FP_{t-i}^- + \alpha_3 D_{2003}$$

$$+ \varphi^+ e^+_{RF,T-1} + \varphi^- e^-_{RF,T-1} + \varepsilon_t \quad (5)$$

where:

ΔFP^{+}_{t-1} and ΔFP^{-}_{t-1} are the farm price increases and decreases respectively;

$e^{+}_{RF,T-1}$ and $e^{-}_{RF,T-1}$ are the positive and negative observations of lagged co-integrating residuals respectively;

β_0, $\beta_{1,i}$, $\beta_{2,i}$, β_3, φ^{+} and φ^{-} are the coefficients to be estimated.

In this paper, we will use the ECM to test only for asymmetry in the speed of price transmission and not in its magnitude. This follows Meyer and von Cramon-Taubadel (2004) who point out that co-integration and ECM are based on the idea of prices being in long-run equilibrium. In fact, prices may drift apart in the long run for reasons unrelated to pure price transmission (e.g. the inclusion of new marketing services), thus it is impossible to test asymmetry in the magnitude of price transmission.

The ECM test for short-run asymmetric speed of price transmission can be represented by equation (3), the same H_0 as for the Houck model. The ECM test for long-run asymmetric speed of price transmission can be represented by the H_0 in equation (6):

$$H_0: \varphi^{+} = \varphi^{-} \qquad (6)$$

A comparison between equations (1) and (5) shows that the ECM nests the Houck model. Capps and Sherwell (2005) argue that if either of the coefficients φ^{+} and φ^{-} are statistically different from zero, the ECM is statistically superior to the Houck model.

3. Threshold Model

Tong (1983) introduced the concept of nonlinear threshold models. In this approach, deviations from the long-run equilibrium between co-integrated price series will only lead to price responses if these deviations exceed a specific threshold level. Meyer (2003) argues that if an ECM is used to estimate price adjustment, there is an implicit assumption that price adjustments induced by deviations from the long-run equilibrium are continuous and a linear function of the magnitude of the deviations from long-run equilibrium. So, even very small deviations from the long-run equilibrium will lead to an

adjustment process on each market, and this is considered unlikely if adjustment costs are present. Threshold models have been used in a number of studies (e.g. Goodwin and Harper, 2000; Serra and Goodwin, 2003; Varra and Goodwin, 2005; Serra *et al.*, 2006; Balcombe *et al.*, 2007). The equations in (7) show a multiple threshold ECM for the Iranian chicken market:

$$\Delta RP_t = \alpha^1_0 + \sum_{i=0}^{L1} \alpha^1_{1,i} \Delta FP_{t-i} + \alpha^1_2 D_{2003} + \varphi^1 e_{RF,t-1} + \varepsilon_t$$

$$if \ e_{RF,T-1} < C_1$$

$$\Delta RP_t = \alpha^2_0 + \sum_{i=0}^{L1} \alpha^2_{1,i} \Delta FP_{t-i} + \alpha^2_2 D_{2003} + \varphi^2 e_{RF,t-1} + \varepsilon_t$$

$$if \ C_1 \le e_{RF,T-1} \le C_2$$

$$\Delta RP_t = \alpha^3_0 + \sum_{i=0}^{L1} \alpha^3_{1,i} \Delta FP_{t-i} + \alpha^3_2 D_{2003} + \varphi^3 e_{RF,t-1} + \varepsilon_t$$

$$if \ C_2 < e_{RF,T-1} \qquad (7)$$

Following Varra and Goodwin (2005) we will use this model to test the following asymmetries in price transmission:

Asymmetry in the speed of price transmission outside the (C_1, C_2) interval;

Asymmetry in the magnitudes before a response is triggered (C_1 and C_2 differ in absolute value)

The estimation procedure for the threshold model used follows Varra and Goodwin (2005) and may be summarized in the following steps.

1- Augmented Dickey-Fuller unit root test and Johansen co-integration test are used to evaluate the time series properties of the data.

2- A co-integrating relationship among the variables is estimated by OLS and the lagged residuals from the co-integrating regression are obtained as the error correction term.

3- A two-dimensional grid search is then conducted to define two thresholds. The procedure searches for the first threshold between 1% and 99% of the largest (in absolute value) negative error correction term. In like fashion, it searches for the second threshold between 1% and 99% of the largest positive error correction term. To choose the thresholds, it needs to search for the minimum of the log of the determinant of the covariance matrix for the residuals. When the optimal threshold is determined, the equations in (7) will be estimated using the threshold values.

RESULTS

The data used in this study are average weekly prices for live chicken (the farm level) and chicken meat (the retail level) over all provinces in Iran for the period October 2002-March 2006. Figure 1 shows the behavior of weekly farm and retail chicken prices in Rials/kg. As may be seen, farm and retail prices have a similar pattern of fluctuations. However, over the period in question, the prices have drifted slightly apart with the farm price rising 17 percent and the retail price rising 19 percent. This resulted in a growth in marketing margin of 22.5 percent over this same period. The reports of the central bank of Iran show that the inflation rates have been 12-15 percent during 2002-2006.

Figure 1: Weekly farm and retail chicken prices
(October 2002-March 2006)
Source of data: Iranian Ministry of Agriculture

Our procedure for testing for asymmetric price transmission involved two preliminary steps:

1- Test for the presence of unit roots in the two price series. This will determine whether the price series need to be first differenced in the estimating equation for price transmission equation;

2- Test for Granger causality of the two price series. This will determine which of the two price series to use as the dependent variable in the estimated price transmission equation.

The basic test for unit roots is the Augmented Dickey-Fuller (ADF) test. The results of the ADF test for both the farm price series and the retail price series are summarized in Table 1. We failed to reject the null hypothesis for a single unit root at both the farm and retail levels and hence conclude the price series are co-integrated.

One problem with the standard ADF test is

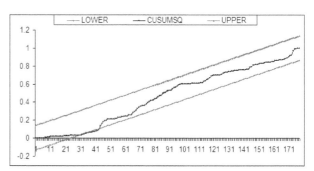

Figure 2: CUSUMQ test for farm and retail prices of chicken

that the test results may be invalidated by the presence of structural breaks in the data series. However, the government's introduction of the price stabilization scheme in May 2003, may have caused exactly that. Hence a CUSUMQ test was used to check for structural breaks in our data. The results of this test are represented in Figure 2 where a structural break is revealed in the 32nd week of the time series, the same time as the government policy intervention.

Table 1: ADF Test for Farm and Retail Chicken Prices

Null Hypothesis	Farm Price		Retail Price	
	Test Statistic	Critical Value*	Test Statistic	Critical Value*
No Trend	-1.83	-2.57	-1.73	-2.57
No Trend, No Constant	1.98	3.78	1.86	3.78
No Trend	-2.58	-3.13	-2.29	-3.13
No Trend, No Constant and Unit Root	2.51	4.03	2.01	4.03
No Trend and Unit Root	3.45	5.34	2.78	5.34

* at the 10 percent significance level

As a result, we applied a modified ADF test proposed by Perron (1990) to test for non-stationarity (a unit root) in the presence of a structural break. We estimated equation (8):

$$Y_t = a_0 + a_1 DU + dDTB + \beta t + \rho Y_{t-1} + \sum_{i=1}^{P} \theta_i \Delta Y_{t-i} + e_t$$

(8)

Where:

Y_t is the times series being tested for non-stationarity;

DU is a dummy variable equal to 1 for observations after the structural break (32nd observation), 0 otherwise;

DTB is a dummy variable equal to 1 for the 33rd observation, 0 otherwise;

With respect to the estimated equation (8), the null hypothesis of the test for the presence of a unit root in Y_t is:

$H_0: \rho = 1$ (9)

where, Perron (1990) has calculated appropriate critical values. The corresponding test statistics for the farm price series and retail price series are -4.17 and -4.04 respectively and the appropriate critical value -4.39 at 1 percent of significance. We thus conclude there is insufficient evidence to reject H_0. The results of the ADF test in the presence of a structural change confirm the results of the standard ADF test.

We then tested for Granger causality between the two price series. Testing that farm price Granger causes retail price yields a highly significant test statistic of 9.1 at 1 percent of significance. However, the converse test that retail price Granger causes farm price yields the insignificant test statistic of 0.22. Thus we set ΔRP as the dependent variable in the price transmission models.

We now turn to test for asymmetric price transmission using the three alternative approaches to estimation: the Houck Model; the Error-Correction Model; and the Threshold Model.

1. The Houck Model

The farm price was first segmented following the Houck procedure. Equation (1) was then estimated using the OLS method. The Ramsey test statistic (F=1.09) suggested that misspecification was not a problem and the Jarque-Bera statistic (9.92) suggested that the residuals are normally distributed. However, the Durbin-Watson (DW) test on this equation suggested the presence of serial correlation. Thus, the equation was re-estimated using the GLS method and the results of this estimation are summarized in Table 2. We used the Akaike Information Criterion (AIC) and Schwarz Information Criterion (SIC) to determine the optimal lag length of farm

Table 2: Houck Model of Farm to Retail Price Transmission

Dependent Variable: ΔRP_t (First Difference of Retail Price) (GLS method)				
Name of Variable	Estimated Coefficient	t statistic	Short-Run Elasticity	Long-Run Elasticity
Intercept	-0.28	-0.17	-	-
ΔFP^-_t (Farm Price Decreases)	0.96	9.7*	0.33	
ΔFP^-_{t-1} (1st Lag of Farm Price Decreases)	0.26	3.58*	0.09	0.42
ΔFP^+_t (Farm Price Increases)	1.32	18.2*	0.52	0.52
D_{2003} (Government Policy)	0.45	0.3	-	
R2		0.83	AIC	5.11
D.W		2.13	SIC	5.21
Price transmission tests		F statistic		Result
Symmetry in Speed of Price Transmission		6.01		Reject
Symmetry in Magnitude of Price Transmission		0.46		Accept

*Significant at 1%

Table 3: Error Correction Model of Farm to Retail Price Transmission

Dependent Variable: ΔRP_t (First Difference of Retail Price) (GLS method)				
Name of Variable	Estimated Coefficient	t statistic	Short-Run Elasticity	Long-Run Elasticity
Intercept	-1.18	-0.68	-	-
ΔFP^-_t (Farm Price Decreases)	0.93	9.96*	0.32	0.36
ΔFP^-_{t-1} (1st Lag of Farm Price Decreases)	0.11	1.54**	0.04	
ΔFP^+_t (Farm Price Increases)	1.29	18.63*	0.51	0.51
D_{2003} (Government Policy)	0.16	0.11	-	-
$e^+_{RF, T-1}$ (Positive Values of Lag Error Term)	-0.29	-3.24*	-	-
$e^-_{RF, T-1}$ (Negative Values of Lag Error Term)	-0.22	-2.91*	-	-
			AIC	4.96
R2		0.86		
D.W		2.04	SIC	5.09
Price transmission tests		F statistic	Result	
Symmetry in Speed of Price Transmission		7.04	Reject	
Symmetry in Price Transmission in Long-Run		0.28	Accept	

*Significant at 1%

price increases and decreases. These two criteria showed that only the first lag of farm price decreases has a significant effect on retail price differences. The high R^2 together with the statistical significance of the estimated regression coefficients confirm the goodness of fit of the model. The small t statistic on the D_{2003} suggests that the government's introduction of the price stabilization policy has not had any significant effect on retail price fluctuations. We estimated Houck Model including Product Dummy variable too but the results showed that the coefficient of this variable is not significant.

A comparison between the coefficients of farm price increases and farm price decreases indicates that retail price is more sensitive to increases than decreases in farm prices. Price transmission elasticities and price transmission tests also confirm that price transmission in the Iranian chicken market is asymmetric and farm price increases transmit more fully and faster than farm price decreases to the retail price.

2. Error Correction Model (ECM)

The ECM as represented by Equation 5 was first estimated using the OLS method. As with the Houck Model, the Ramsey test statistic (F=2.3) suggested no evidence of misspecification and the Jarque-Bera statistic (12.68) suggested that the residuals were normally distributed. However, once again the Durbin-Watson test confirmed the presence of serial correlation. Hence the ECM was re-estimated using the GLS method and the results are summarized in Table 3. The high R^2 together with the statistical significance of the estimated regression coefficients confirm the goodness of fit of the model. As expected the coefficients of farm price increases and decreases have positive sign indicating a positive relationship between farm and retail prices. Further, as expected, the coefficients of the positive and negative values of the lag of error correction term have a negative sign indicating that any deviation from long-run equilibrium between farm price and retail price in one period will tend to be compensated for in the next.

As in the estimated Houck model, the AIC

and SIC confirm the statistical significance of a first lag of farm price decreases. The coefficients and corresponding elasticities of ΔFP^-_t, ΔFP^-_{t-1} and ΔFP^+_t show that farm price increases transmit to the retail level more fully and quickly than farm price decreases. The coefficients of $e^+_{RF,T-1}$ and $e^-_{RF,T-1}$ suggest that positive deviations from long-run equilibrium will correct more quickly than negative deviations but this difference is not significant and the null hypothesis of equality between them that is the test for symmetry in price transmission in long-run is accepted.

The ECM differs from the Houck model in its inclusion of $e^+_{RF,T-1}$ and $e^-_{RF,T-1}$ as additional explanatory variables. Since both were found to be significant, we may conclude that the ECM is superior to the Houck model. The F test for model selection (F=184.3) confirms the superiority of the ECM to the Houck model. However, both the Houck model and the ECM lead to a rejection of the null hypothesis of symmetric price transmission in the Iranian chicken market.

3. Threshold Model

To analyze price transmission behavior in the Iranian chicken market using Threshold approach, we first calculated threshold values using the grid search procedure to find the minimum value of the log of the determinant of the covariance matrix for the residuals as explained in the Methodology above. We found two thresholds (3.4) and (-5.4) and estimated the equations in (7) using the OLS method. The results of our estimation are presented in table 4. The high R^2 together with the statistical significance of the estimated regression coefficients confirm the goodness of fit of the model. The Durbin-Watson statistic suggests that serial correlation is not evident in this model. As expected, the coefficient of the error correction term in the second regime $(-5.4 \leq e_{RF,T-1} \leq 3.4)$ is not significant.

The existence of two thresholds suggests that

deviations in the positive and negative directions must reach different magnitudes before a response is triggered and hence price transmission in the Iranian chicken market is asymmetric in magnitude. A comparison between the coefficients of $e_{RF,T-1}$ in the first and third regimes confirms that there is asymmetry with respect to the speed of price transmission. As with the ECM and the Houck Model, D_{2003} is not a significant explanatory variable. This suggests the Iranian government's price stabilization policy has not been successful in decreasing retail chicken price fluctuations, at least at the retail level.

The Threshold Model confirms the results of the ECM and the Houck model in rejecting the null hypothesis of symmetric with respect to the speed of price transmission in the Iranian chicken market in short-run. However, the Threshold Model may represent an improved specification as it allows for the existence of thresholds of varying magnitudes.

CONCLUSIONS

In this article we used three alternative approaches to analyze the existence of asymmetric price transmission between the farm and retail levels in the Iranian chicken market. The three approaches involved using the Houck Model, the Error Correction Model (ECM) and the Threshold Model. The analysis suggests that farm prices (Granger) cause retail prices and all three approaches suggest the price transmission process is asymmetric in short-run. Statistical tests show that the ECM is superior to the Houck model and the existence of thresholds suggests that the Threshold model is superior to the ECM. Price transmission elasticities for farm price increases were found to be larger than for farm price decreases suggesting that the speed of price transmission is greater when prices are rising than when prices are falling in short-run. This is a positive asymmetric price transmission and is beneficial for marketing agents. On the other hand, results of ECM and Threshold model show that if the retail price is above its equilibrium, this deviation is corrected faster than if the retail price is below its equilibrium. This is a negative asymmetric price

Table 4: Threshold Model of Farm to Retail Price Transmission

Dependent Variable: (First Difference of Retail Price)		
Name of Variable	Estimated Parameter	t statistic
Intercept	19.95	4.95*
ΔFP_t (1st Difference of Farm Price)	1.27	23.57*
$e_{RF, t-1}$ (Lag Error Term)	-0.69	-6.41*
D_{2003} (Government Policy)	-1.03	-.29
R2	0.9	
D.W	2.24	
AIC	5.14	
SIC	5.27	
n	75	

$$3.4 \leq e_{RF, t-1}$$

Name of Variable	Estimated Parameter	t statistic
Intercept	19.95	4.95*
ΔFP_t (1st Difference of Farm Price)	1.27	23.57*
$e_{RF, t-1}$ (Lag Error Term)	-0.69	-6.41*
D_{2003} (Government Policy)	-1.03	-.29
R2	0.9	
D.W	2.24	
AIC	5.14	
SIC	5.27	
n	75	

$$-5.4 \leq e_{RF, t-1} \leq 3.4$$

Name of Variable	Estimated Parameter	t statistic
Intercept	-9.71	-2.66**
ΔFP_t (1st Difference of Farm Price)	1.33	25.65*
$e_{RF, t-1}$ (Lag Error Term)	-0.43	-2.99**
D_{2003} (Government Policy)	-3.65	-1.14
R2	0.9	
D.W	2.24	
AIC	5.14	
SIC	5.27	
n	75	

$$e_{RF, t-1} \leq -5.4$$

*Significant at 1%   **Significant at 5%

transmission in long-run and is beneficial for consumers but this asymmetry is not statistically significant. Results of our estimation also suggest that the introduction of the Government's price stabilization policy has not been effective in decreasing price fluctuations, at least at the retail level. We also expected symmetric price transmission in presence of the Government's price stabilization policy but this hypothesis is rejected.

We believe that asymmetric price transmission in Iranian chicken market is the result of high inflation rates, and a non-competitive slaughtering industry. High inflation rates conduct to the positive asymmetric price transmission in two ways. First; inflation leads the consumers to expect continual price increases. Aguiar and Santana (2002) have found an evidence of asymmetric price transmission in presence of inflation in Brazil too. Second; as Ball and Mankiw (1994) mentioned, in presence of positive trend inflation rates, different adjustment costs lead to asymmetric price transmission. "In presence of positive inflation trend, positive shocks to firms' desired prices trigger greater adjustment than do negative shocks of the same size. Indeed, inflation causes firms' relative prices to decline automatically between adjustments. When a firm wants a lower relative price, it need not pay the adjustment cost, because inflation does much of the work. By contrast, a positive shock means that the firm's desired relative price rises while its actual relative price is falling, creating a large gap between desired and actual prices. As a result, positive shocks are more likely to induce price adjustment than are negative shocks, and the positive adjustment that occur are larger than the negative adjustment."

We believe that one of the reasons for asymmetric price transmission in Iranian chicken market is non-competitive structure and existence of market power in slaughtering industry. Thus, seeking for generating more competitive markets will help to have symmetric price transmission in Iranian chicken market and consumers will gain from positive welfare effects of symmetric price transmission.

REFERENCES

1- Aguiar, D. R. D. & J. A., Santana, (2002) "Asymmetry in Farm to Retail Price Transmission: Evidence from Brazil". Agribusiness, Vol. 18(1), pp. 37-48.

2- Bailey, D. & B.W., Brorsen, (1989) "Price Asymmetry in Spatial Fed Cattle Markets" Western Journal of Agricultural Economics, Vol. 14(2), pp. 246-252.

3- Balcombe, K., D., Bailey, and J. Brooks, (2007), "Threshold Effects in Price Transmission: the Case of Brazilian Wheat, Maize and Soya Prices", American Journal of Agricultural Economics, Vol. 89, pp. 308-323.

4- Balke, N. S. & T. B., Fomby, (1997) "Threshold Cointegration". International Economic Review, Vol. 38, pp. 627-645.

5- Ball, L. & N. G., Mankiw, (1994) "Asymmetric Price Adjustment and Economic Fluctuations" Economics Journal, Vol. 104, pp. 247-261.

6- Borenstein, S., A. C., Cameron, & R., Gilbert, (1997) "Do Gasoline Prices Respond Asymmetrically to Crude Oil Price Changes?" Quarterly Journal of Economics, Vol. 112, pp. 305-339.

7- Boyd, M. S. & B. W., Brorsen, (1988) "Price Asymmetry in the U.S. Pork Marketing Channel" North Central Journal of Agricultural Economics, Vol. 10, pp. 103-109.

8- Capps, J. O. & P., Sherwell, (2005) "Spatial Asymmetry in Farm-Retail Price Transmission Associated with Fluid Milk Products" Selected Paper prepared for presentation at the American Agricultural Economics Association Annual Meeting, Providence, Rhode Island, July 24-27, 2005.

9- Goodwin, B. K. & D. C., Harper, (2000) "Price Transmission, Threshold Behavior and Asymmetric Adjustment in the U.S. Pork Sector" Journal of Agricultural and Applied Economics, Vol. 32, pp. 543-553.

10- Hahn, W. F., (1990) "Price Transmission Asymmetry in Pork and Beef Markets" The Journal of Agricultural Economics Research, Vol. 42(4), pp. 21-30.

11- Houck, J. P., (1977) "An Approach to Specifying and Estimating Nonreversible Functions" American Journal of Agricultural Economics, Vol. 59(3), pp. 570-572.

12- Kinnucan, H. W. & O. D., Forker, (1987) "Asymmetry in Farm-Retail Price Transmission for Major Dairy products" American Journal of Agricultural Economics, Vol. 69(2), pp. 285-292.

13- Meyer, J. & S., Von Cramon-Taubadel, (2004) "Asymmetric Price Transmission: A Survey" Journal of Agricultural Economics, Vol. 55(3), pp. 581-611.

14- Meyer, J., (2003) "Measuring market integration in the presence of transaction costs: A threshold vector error correction approach" Contributed Paper selected for presentation at the 25th International Conference of Agricultural Economists, August 16-22, 2003,Durban, South Africa.

15- Mohanty, S., E. W. F., Peterson & N.C., Kruse, (1995) "Price Asymmetry in the International Wheat Market" Canadian Journal of Agricultural Economics, Vol. 43, pp. 355-366.

16- Perron, P., (1990) "Testing for a Unit Root in a Time Series with a Changing Mean" Journal of Business and Econometrics, Vol. 33, pp. 311-40

17- Scholnick, B., (1996) "Asymmetric Adjustment of Commercial Bank Interest Rates: Evidence from Malaysia and Singapore". Journal of International Money and Finance, Vol. 15, pp. 485-496.

18- Serra, T. & B. K., Goodwin, (2003) "Price transmission and Asymmetric Adjustment in the Spanish Dairy Sector". Applied Economics, Vol. 35, pp. 1889–1899.

19- Serra, T. Gill & K., Goodwin. (2006). "Local Polynomial Fitting and Spatial Price Relationships: Price Transmission in EU Pork Markets". European Review of Agricutural Economics, Vol. 33(3), pp. 415-436.

20- Tong, H., (1983) "Threshold Models in Non-Linear Time Series Analysis" New York, Springer-Verlag.

21- V. Cramon-Taubadel, S. & J. P., Loy, (1996) "Price Asymmetry in the International Wheat Market: Comment" Canadian Journal of Agricultural Economics, Vol. 44, pp. 311-317.

22- Varra, P. & Goodwin, B. K., (2005) "Analysis of Price Transmission along the Food Chain" OECD Food, Agricultural and Fisheries Working Paper, No. 3, OECD Paris.

23- Ward, R. W., (1982) "Asymmetry in Retail, Wholesale and Shipping Point Pricing for Fresh Vegetables" American Journal of Agricultural Economics, Vol. 62(2), pp. 205-212.

24- Wolffram, R., (1971) "Positivistic Measure of Aggregate Supply Elasticities: Some New Approaches – Some Critical Notes" American Journal of Agricultural Economics, Vol. 53(2), pp. 356-359.

13

How Should Rural Women's Enterprises Be Developed and Promoted?

Reza Movahedi [1]*, Masoud Samian[2], Khalil Mirzai [2] and Adel Esmaili Saloomahalleh [3]

Abstract

This study was a qualitative research which was done by a case study approach. The samples were chosen from rural development experts of Hamedan Province during a snowball sampling process. The data were collected by the semi-structured interviews. The data collection was done by face-to-face interviews, after the interview questions guide was determined. The theoretical saturation was attained by interviewing 25 respondents and at this point the interview process was laid off. The qualitative analysis of interviews was done by a content analysis method. The results showed that in order to develop and promote the rural women enterprises not only financial, domestic, managerial, educational and cultural supports are needed but some approaches like culturalization and educational programs are vital too.

Keywords:
Women Enterprises, Business, Rural Development, Qualitative Research

[1] *Assistant Professor Agricultural Extension and Education Department, Bu-Ali- Sina University, Hamedan, Iran.*

[2] *M.Sc. Student, Agricultural Extension and Education Department, Bu-Ali Sina University, Hamedan, Iran.*

[3] *M.Sc. Student, English Language Education Department, Islamic Azad University, Takestan Branch, Takestan, Iran.*

* *Corresponding author's email: r.movahedi@basu.ac.ir*

INTRODUCTION

The changing from the present system to a sustainable system could be accomplished by the participation of all individuals. Hence, the role of the women, which constitutes half of the population of the country, must be taken into account (Veisi and Badsar, 2005). Identifying and trying to use all of the human capacity in rural communities is central to achieving sustainable rural development goals. This approach shows the importance of identifying the role of rural women, and recognizing their status as half of the rural population, and highlights the necessity of utilizing the women's capabilities. Certainly, rural women can be a viable force for change and a potential source for growth of the rural economy, including an increase in employment (Movahedi and Yaghoubi-Farani, 2012).

The situation of rural women and their activities in Iran indicates that despite the significant role of women in economic, social and cultural activities, the employment of rural women is not in a desirable situation (Fallah-Jelodar *et al.,* 2007). Rural women have performed lots of social and economic indoor roles as well as outdoor ones. But, unfortunately, their actual position and value is not determined. They have been excluded from rural development plans; and, they have been only attributed to babysitting, feeding and etc. These issues caused the designers and programmers to neglect the particular potential of rural women (Mansourabadi and Karami, 2006).

The women should try to find some procedures in order to develop their thoughts and attitudes. They can strengthen their ability and competence by the making of the best use of participating actively in different courses; and, by this way, they can improve their status in the society. The promotion of women's position in the society is up to utilizing the opportunities, fair and equal circumstances for women. One of the most essential factors that cause poverty and penury for women is illiteracy and lack of knowledge about social, economic and even hygienic status. To be educated and to utilize educational facilities provide an opportunity for women that enable them to promote their family and society by recognizing their biological and social status. According to statistics, the higher the women's educational level, the healthier individuals would be (UNESCO, 2004). If the effective role of women in production had been ignored, the economic development of any society would have encountered some dilemma; because, the financial management role of women in families deeply affects family economy, and in upper levels, society economy (Salahshour, 2003).

Traditionally, Iranian rural women have done some activities as: transplanting vegetables, weeding beet and cotton, pounding and grinding grains, sifting crops, well-setting bestial affairs and meat drying for the cold season utilization (Shahbazi, 2003). Although rural women are occupied with house-keeping, husbandry, breeding children and they also are involved in agricultural activities, their economic-social dignity is not deserved for their endeavor and activity; and finally, the illusion of rural woman in our country shows a hard-working, sufficient, patient and strong picture (Alayirahmani, 1996).

Research Literature

According to some development theories, it is believed that all the social dimensions, parties and areas must be developed. Based on this theory, development experts believe that development in all the social aspects would be somehow impossible when the women are ignored. Therefore, women must be brought in development plans and they must become capable to participate in economic and political structure of the society. In order to determine the women status in development, it is primarily essential to investigate the human role and status in the development process.

In 1980, spreading sustainable development had caused the women role to be entered into a new phase. The advocates of sustainable human development believe that "sustainable human development" occurs only by providing equality and parity between sexes. Whenever the women are out of development cycle, the development would be incapable, though. The existence of equality between sexes is essential in all the life dimensions in sustainable development (Malek-mohamadi and Hosseininia, 1999).

The sustainable development attitude has located women in the center of development definitions and considering the women position in development has interned it into a new phase. On the other side, the agricultural executive policies must be continuously revised in such a way that increases the participation of rural women in production. The particular developmental plans, with regard to security circumstances for rural women, should increase the financial facilities of this active group, while their job pressure and hardship should be decreased.

Other supportive plans should consider all rural girls and women's needs in a systemic and comprehensive way, and this empowers them not only to help food security of family but also to implement health care. Moreover, the continuous revising of agricultural policies in order to meet the rural women needs and determining their priorities are of specified development tools and it increases agricultural products. The agricultural development projects will promote the rural general awareness and also urban life, if it is designed in a framework that increases the rural women empowerment.

Some achievements are related to extending the women participation in political, economic, social and cultural fields and accentuating their role in various situations has led to the creation of specific institutions. For example, in Iran, as a developing country, the most significant of such institutions are: the Women Socio-Cultural Council, the Women and Juvenile Specific Committee in the System Expedient Recognition Convention, the Women Participation Center in Presidency Institution, the Women Affairs Province Commission, the Women Commission in Islamic Council Parliament, the Women Affair Office in Judicature and also the other institutions that are doing some activities indirectly for women such as Imam Khomeini Relief Committee, Agriculture Ministry, Literacy Movement. Regarding the importance of the women participation and exploiting their competence for the recognition and resolving their problems, the organized and systematic planning in different fields is needed. By constituting special groups and institutions of women, particularly the rural women institutions, undoubtedly, it can form the woman's attendance in different fields and it removes the undeserved understanding of the women's positions and rights. On the other hand, statistically, the rural women population in Iran is about 12 million people; and 42 percent of rural women population are between 15 to 24 and 13.11 percent are adult and finally 9.4 percent are 65 and more (Shafaghi, 2001).

Generally, the enterprises which are related to women follow special plans and purposes. These goals can be defined as follow: to make an attempt to establish equality between women and men, to try to gain experiences and knowledge about women rights, to try to provide a situation for women to participate in conferences and seminars, to create secondary commissions to consider the women issues. To do this, the coordinator systems of the women NGO (non-governmental organizations) are constructed (Ahmadiyan, 1999). So, the rural woman enterprises can be both local and self-constructed that can be managed non-governmentally and it may be in the format of designs and projects that are supported by governmental organizations and it has an active role in all implementations and credit stages. As mentioned above, one of the most important elements of rural development is to make the women capable in managing fields and promoting their function in the same field, and to make decision and to accomplish in family and society. The participation value of the rural women would be increased in product and economy of the country by performing policies and codifying guiding plans by the use of educational, promoter and entrepreneurship plans. This would be possible by the creation and promotion of human enterprises and institutions in rural areas (Mafi and Mousavi, 2004).

Williamson (2005) believes that, in rural sustainable development, coordinating the activities would be effective when it is compatible with nature not when it is against it. The rural areas in under-developed [third-world] countries have a prominent role in the development of a country by allocating a large portion of Gross National Products (GNP), occupation, providing food needs and habitat of a large population of people; as, one of the most effective approach for a strong and sustainable development of these areas would be making the use of partnership and competence of all individuals especially the rural to develop the rural areas in all dimensions (Dorzinia and Amiri, 2005).

The World Bank (2000) considers the rural societies participation as an essential path in accomplishing the rural development priorities in order to making the best use of the present sources in rural areas. Warren-Smith and Jackson (2004) underline the importance of rural social participation as both a big step in establishing democracy between social organizations and the creation of self-aiding sense between the rural. The commitment and persistence of these local enterprises forced the government to assign some of its heavy responsibilities to these local organizations while the government believes in their capability and effectiveness of these NGOs and their position. The NGOs define the human participation in society by the use of their medium [intermediate] role between the individuals and the government (Gholfi, 2003).

Stenseke (2009) mentions the rural participation, in the formation of rural enterprises, as an approach whose duty is to support the cultural values of an area -regarding the individuals participation and also success of enterprises in rural areas; there are two points that must be taken into consideration: first, the individuals should be encouraged to participate, because they have a key role in rural areas development; second, to provide some opportunities for the rural demands because the creation of human enterprises can be considered as a bridge that connects the rural needs to responsible men (Coelho and Favareto, 2008).

MATERIALS AND METHODS

The present study was a qualitative one which was done by the use of case study. The samples were chosen by considering the topic recognition criterion from rural development experts of college and the administration part of Jihad-Agriculture Organization of Hamedan Province. The main criteria of selecting materials in qualitative researches are to be experienced and to be familiar with the studied phenomenon. The data collection method was the semi-structured interviews. First, the topic concept was investigated and by the use of it four principals were clarified as the research questions; and, interview questions guide were specified based on these questions. These four main research questions were as follow:

1- What are the roles and functions of women rural enterprises?

2- What types of supports, facilities and circumstances are necessary to operate rural women enterprises?

3- What are the barriers and limitations of rural women in rural women enterprises?

4- What are the solutions and approaches of promotion and development of rural women enterprises?

The data collection was done by face-to-face interviews, after the interview questions guide was determined. The number of samples was not clear at the beginning, but, as soon as the interview was started, the other key samples were defined by snowball sampling method and the data collection process was kept on till the samples theoretical saturation step. The theoretical saturation was attained by interviewing of 25 samples and at this point the interview process was laid off. The qualitative sample analysis of interviews was done by the content analysis method. To do this, first, the data were coded and summarized separately and finally the common topics were reported. Regarding that the number of samples was more than 20 cases, according to researches, the common topics were mentioned in the form of rehearsal frequency and percentage. So, in this paper the common topics were reported in the form of rehearsal frequency and percentage.

RESULTS AND DISCUSSION

The expected experts are between 32 to 49 years of age. The average age of individuals is 38. Forty percent of samples were male and 60

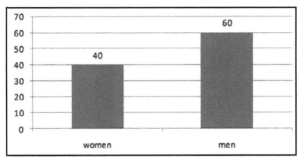

Figure 1: The percentage of gender samples

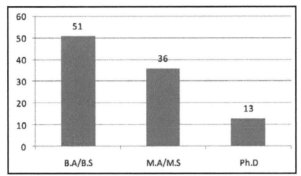

Figure 2: The percentage of education degree of samples

percent were female. Their job experience average was 8 years. The degree of interviewed sample were as follow: 51 percent B.S., 36 percent M.S. and 13 percent Ph.D. (graph 1 and 2)

Interview question 1: Roles and function of rural women enterprises

The experts of rural development issues were interviewed and their responses were collected. Totally, different responses were collected which its data analysis is depicted in the following table. The most significant roles and functions of rural women development, as illustrated in table 1, can be categorized in economic, social, individual, domestic, juridical, educational and managerial functions.

Interview question 2: Supports, facilities and circumstances to develop rural women enterprises

The interviewers were asked to state their opinions openly about the essential support of implementing rural women enterprises. The results show that, as illustrated in table 2, lawful and financial, domestic, managerial, educational and cultural supports are of upmost importance which is necessary for the women enterprise development from the respondents' point of view.

Here, some comments are quoted directly:

"Regarding to the activity and enterprise type, bank credits and services must be available for women. The women who are cultivating mushroom, as an example, must be exposed to a proper place and bank conveniences to do so."

"The location of the women enterprise must be such a proper and domestic place; hence, their families would let them participate in those enterprises. This place should be designed in a way which the mothers who have children could visit their children as easy as possible; for instance, such places like kindergarten must be available to take care of children; so, the mother can participate in the enterprises."

Table 1: Roles and functions of rural women enterprises

Priority	Common topics	frequency	The percent of each element separately	The percent of each element related to others
1	Economic function (earning development, income increasing)	22	88	18.2
2	Social function (entrepreneurship, participation and women social relation development)	21	84	17.4
3	Individual function (self-confidence, motivation and individual creativity increase)	19	76	15.7
4	Family function (family health, mental and physical health, discrimination removal)	17	68	14
5	Juridical function (familiarity with the women`s right, financial rules)	16	64	13.2
6	Technical and educational function (preoccupation and modern knowledge familiarity)	14	56	11.6
7	Managerial function (decision-making and planning power promotion)	12	48	9.9

Table 2: Essential supports, facilities and circumstances for implementation of rural women enterprises

Priority	Common topics	frequency	The percent of each element separately	The percent of each element related to others
1	Financial and juridical supports (bank credits)	24	96	23.8
2	Family supports	22	88	21.8
3	The enterprises planning and management supports (proper place and time, providing suitable situation)	21	84	20.8
4	Technical and educational supports	18	72	17.8
5	Medium and cultural supports	16	64	15.8

Interview question 3: The solutions for the development and promotion the rural women enterprises

The respondents were asked to state their opinions about solutions and approaches of rural women enterprises development and promotion. After collecting the responses, different issues were coded by content analysis and then they were classified in common topic as illustrated in table 3.

Interview question 4: Barriers and limitations of development the rural women enterprises

Here, the respondents' opinions were analyzed and the results were summarized in table 4. As the results show, the cultural, individual and domestic, economic, social barriers are identified as the most important barriers of development of the women enterprises.

Here, another comment is stated:

"To me, the first thing about the rural women is the matter of their family thought; maybe their husband, father-in-law, brother and daughters disagree with their activity. The large distance of the location of rural women enterprise may cause lots of problems; and, rural women are not able to travel to far places. Culture and wrong beliefs is really widespread in rural areas. In Kurdish rural areas, affluent women usually do not work; and, their families are ashamed of women's working, except when they have to do so or when they are poor.

CONCLUSIONS AND RECOMMENDATIONS

The results indicate that rural women enterprises possess different roles and function which are considered as an important part of rural development. The most vital roles and functions that are mentioned in this paper are as follow: economic, social, individual, domestic, juridical, educational and managerial functions.

The results also show that in order to develop and promote the rural women enterprises not only implementing lawful, financial, domestic, managerial, education and cultural is needed but some approaches like culturalization and implementing plans to empower women, the continuous implementation of educational and counseling plans, encouraging successful women enterprises are vital too.

Table 3: The solutions and approaches of the development and promotion the rural women enterprises

Priority	Common topics	frequency	The percent of each element separately	The percent of each element related to others
1	Proper culturalization and suiting (in individual, family, social and economic dimension)	23	92	30.3
2	Implementing plans to make the women more competent (political, managerial, official)	21	84	27.6
3	The continuous implementation of counseling and educational plans (introducing enterprises, familiarity, informing)	18	72	23.7
4	Encouragement and introducing successful enterprises	14	56	18.4

Table4: barriers and limitations of rural women enterprises

Priority	Common topics	frequency	The percent of each element separately	The percent of each element related to others
1	Cultural barriers (wrong beliefs in rural areas, lack of freedom)	24	96	27.9
2	Family and individual barriers (lack of self-confidence and physical power, lack of family permit to work)	22	88	25.6
3	Economic barriers (lack of loan, and bank credits, lack of the women`s financial dependency)	21	84	24.4
4	Social barriers (lack of society admission, having different responsibilities)	19	76	22.1

The major limitation and barriers of rural women in developing such enterprises are: cultural barriers particularly the existence of the dominant thought in rural areas, lack of enough information and the women weakness in education and illiteracy, their semi-literacy, economic barriers specially financial problems their economic dependency to their husbands and social barriers.

Therefore, it can be said that the rural women`s participation is a vital and essential issue in implementation of different developmental and agricultural rural activities. Hence, the rural development would be impossible without the women participation. Also, rural enterprises, as a supporting and cooperating and motivating element, can have an important role in making the rural women more competent. For this, it is suggested that these enterprises must be invigorated by the use of supportive, juridical and financial policies and the rural women enterprises development make the women more confident and competent.

REFERENCES

1- Ahmadiyan, M. (1999). Women's organizations and associations underlying cultural, social development. Nashriyeh Neshat. June,1999. (In Persian).

2- Alayirahmani, F. (1996). Hidden Role of rural women in economic development . Simayeh zan dar jameah Conference Proceedings. Tehran: Alzahra University. (In Persian).

3- Coelho, V. S. and Favareto, A. (2008). Questioning the relationship between participation and development . World Development. 36(12): 214- 223.

4- Dorzinia , M . and Amiri , Z. (2004). Review the role and status of NGOs in rural development. Retrieved: 11, 02, 2012, Available: www. socialogy-oofiran.cm.

5- Fallah-Jelodar, R., Hosseini, S.J.F. and Mirdamadi, S.M. (2007) 'Factor affecting on success of rural women's entrepreneurship in north of Iran', Quarterly Journal of Rural and Development, 10(4): 87–115. (In Persian).

6- Gholfi , M . (2004). Networking approach in the non-inkwell. Mahnameh Salehin-e- roosta, 266. (In Persian).

7- Hosein, A.A. (2003). Empowerment women in Autonomy and Decentralization process: experience an Indonesian. Retrieved: 10,11, 2012, Available: www.newyork.doc/es/ah/5.03.

8- Jafarzadeh, F. (2001). Word for Time: Tvata in speech and action. A woman. Retrieved: 11,12, 2012, Available:www.wikipedia.org.

9- Leys , A . Rozariyo , T. and Rozariyo, D. (1994) . Gender and Development . Interpreter : Yousefiyan. Tehran: Banoo Press.

10- Mafi , F. and Mousavi, F. (2003). Aperture to the green effort. Office of women affairs, Ministry of Jihad-Agriculture. (In Persian).

11- Malekmohamadi , A . and Hossininia , GH. (1999). Attitude of rural women in agricultural activities, livestock, food, environment and rural extension programs. Jihad Monthly Magazine. 19: 220-221. (In Persian).

12- Mansourabadi, A. and Karami, A. (2006). Consequences on economic development, social and cultural rights of rural women: case study in the Fars. Human and social journal of Shiraz University. 24(2): 107-128.

13- Movahedi, R. and Yaghoubi-Farani, A. (2012). Analysis of the barriers and limitations for the development of rural women's entrepreneurship. Int. J. Entrepreneurship and Small Business, 15(4): 469–487.

14- Polkinghorne, D. E. (1989). Phenomenological Research Methods Experimental Phenomenological Perspectives in Psychology (Valle, R.S. and Halling, S. Eds).New York, Plenum Press.

15- Shafaghi, M . (2001). Efficient use of capacity planning for women must be comprehensive. Majaleh Irandokht. July, 2001. (In Persian).

16- Shahbazi, E. (2003).Villagers scientist. Tehran: Moaseseh Shaghayegh Roosta.

17- Tavanaelmi, F. (2000). From rice paddies to the foot of the rug. Khorasan newspaper, 102.

18- UNESCO, (2004).What Work in Learning Programmes for Young and Adult Women? Retrieved: 10, 23, 2012, Available: www.unesco.com.

19- Veysi , H . and Badsar, M. (2005) . Factors influencing the social role of women in rural activities: case study in Kurdistan. Quarterly Journal of Rural and Development.8(4): 19-42.

20- Warren-Smith, I. and Jackson, C. (2004). Women creating wealth through rural enterprise. International Journal of Entrepreneurial Behaviour & Research, 10(6): 369 – 383.

21- World Bank (2000). Overcoming Rural Poverty. Conference Draft. Washington, D.C.

22- Wretz, F.J. and Van Zuuren F.J. (1987). Qualitative Research Educational Considerations: Advances in Qualitative Psychology, Themes and Variations. North America, Berwyn.

14

Effectiveness of Crop Advisory Services in Aurangabad District of Maharashtra in India

Bhaskar Pant [1]*, Alpa Rathi [2] and Anshul Rathi [3]

The project was undertaken to study the evaluation of effectiveness of crop advisory services and suggested measures for filling the gap in Aurangabad district of Maharashtra in India. The survey was carried out in 2010. The data was collected with the help of a specifically designed and pre-tested questionnaire. The project carried out in catchment area of advisory services has given substantial insight on the current status of different dimensions of advisory services running in Aurangabad and also recommends strategies to make advisory services accessible to all. The farmer's willingness to pay assumes a key role in determining the success of a cost-recovery strategy. During the study it was interesting to note that of all the 115 respondents 46.67% agreed that their critical need was supply of inputs followed by credit purchase on which the advisory services provider should focus. The dissemination channels were not utilized properly. The results of correlation study indicate that the recommendations by the advisory services and the results after advise have a positive correlation with increase in yield showing the effectiveness of these crop advisory services. The results of multivariate regression indicated that the cropping and harvesting method, credit access, input supply linkage, insurance, age, education and interaction with other farmers have the main role in showing the variations of attitude to adoption of the advisory services.

Keywords:
Channels, Recommendations, Critical need, Yield, Effectiveness, Multivariate regression

[1] *Assistant Manager, Axis Bank Ltd.*
[2] *Deputy Manager, National Commodity & Derivatives Exchange Ltd.*
[3] *Senior Research Scholar, College of Technology and Engineering*
* *Corresponding author's email: bhaskarpnt@gmail.com*

INTRODUCTION

Agriculture currently accounts for 14.6 % of the national GDP (Economic Survey, 2010-11) and provides employment to about 52% of the total work force in India. Agriculture is the backbone of Indian Economy. In the era of globalization and phenomenal technological progress, Indian agriculture faces the twin challenges of meeting the rising demand of food for increasing population in a sustainable manner and making the best use of available resources and technology for enhancing the production and productivity of agricultural sector. The national average yields of most commodities are low. In many areas there are limits to achievable increase in productivity, unless appropriate institutions that can help farmers to access information, inputs and services are strengthened, and joint action for natural resources management, marketing and processing are promoted. Agricultural extension services (in the public as well as private sector) need to play a much larger role in assisting farmers in meeting the above challenges. Extension services are an important element within the array of market and nonmarket entities and agents that provide human capital-enhancing inputs, as well as flows of information that can improve farmers' and other rural peoples' welfare. The goals of extension includes transferring of knowledge from researchers to farmers, advising farmers in their decision making and educating farmers on how to make better decisions, enabling farmers to clarify their own goals and possibilities, and stimulating desirable agricultural developments (van der Ban & Hawkins, 1996). As Schultz (1975) has argued, agriculture-specific human capital is important in improving farm yields in a changing environment because it enhances resource allocation abilities of farmers. Sustainability of an extension service depends crucially on its ability to provide benefits and generate support from internal and external stakeholders (Gustafson, 1994).Yet development scholars and practitioners have generally concluded that the performance of extension services in developing countries has been disappointing. (Rivera et al., 2001).

Public extension alone can't meet the specific need of various regions and different classes of farmers and policy to promote private and community driven extension to complement, supplement, work in partnership and even substitute for public extension is required (Department of Agriculture and Co-operation, 2000). The World Development Report 2008 (World Bank, 2007) emphasizes agricultural extension as an important development intervention for: (1) increasing the growth potential of the agricultural sector in the light of rising demand- and supply-side pressures, and (2) promoting sustainable, inclusive, and pro-poor agricultural and hence economic development.

The improvement in agricultural productivity requires demand-driven and farmer-accountable, need-specific, purpose-specific, and target-specific extension services. In order to use extension approaches that best fit a particular situation, the agricultural extension system has to be sufficiently flexible to accommodate the different options. To this end, the recent agricultural-sector reforms have been geared towards creating a demand-driven, broad-based, and holistic agricultural extension system (Sulaiman & Hall, 2002, 2004; India Planning Commission, 2005).

The services that extension systems need to provide to cover a much wider agenda than the traditional technology transfer function, includes:

1- linking farmers to domestic and international markets, reducing their vulnerability and enhancing the voice of the rural poor (Farrington et al., 2002);

2- advising on and promoting environmental conservation (Alex et al., 2002);

3- advising on farm and small rural business enterprise development and non-farm employment (Rivera et al., 2002);

4- augmenting technology transfer with services relating to both input and output markets (Neuchatel Group, 2002); and

5- contributing to capacity development through training, strengthening the innovation process, building linkages between farmers and other agencies, and helping to strengthen farmers' bargaining position through appropriate institutional and organizational development (Sulaiman & Hall, 2002).

While there is limited evidence on income gains, a recent case study by (Goyal, 2010) suggests that the presence of e-Choupals in Madhya Pradesh in India has increased the average price in government-regulated wholesale agricultural markets in a district by 1-3 percent, and raised farmers' net profits by 33 percent. This e-Choupal business model has eliminated the middlemen from the transaction, which has made the system attractive for both the farmers and the company.

Evaluating the results of the assessment study of gender – related reforms, Danida (2002) concluded that all extension services projects have improved the economic status of trained women and contributed to poverty reduction and reflected the corresponding effect of (1) higher crop yields (2) savings on the use of chemical fertilizer, and (3) higher agricultural productivity through improvements in agricultural practices.

In many developing countries, lack and shortage of relevant and appropriate technologies to improve productivity is a major constraint confronting the extension service, a problem which is more serious in rainfed, resource poor environments (Axinn, 1988; Purcell & Anderson, 1997). Part of the reason for the lack and shortage of appropriate technologies is the weak linkage between research, extension and farmers. To alleviate the aforementioned generic problems of extension, a range of institutional arrangements have been tried, including improvements in extension management, decentralization, and commodity-focused approach, fee-for-service public provision, institutional pluralism, empowerment and participatory approaches, privatization, service contracting, and inter-connecting rural people and use of appropriate media (Anandajayasekeram et al., 2005)

MATERIALS AND METHODS

The main objectives of the study:

1- to evaluate advisory services by knowing its effectiveness, reach and comparative advantages to farmers in catchment area of advisory service providers.

2- to explore gaps for slow adoption of the advisory services and suggest strategy for filling them.

The research conducted was of evaluation type research. Primary data was collected for the purpose of study by adopting survey method. A well structured questionnaire with both closed and open-ended questions was prepared and administered among the farmers living in catchment area of advisory services. Open ended questions were used to evaluate performance of field visit & technical advice and evaluate results after advice, while closed ended questions were used to evaluate customer satisfaction level. A 5-point likert's scale has been used to measure the performance field visit & technical advice of advisory services. Open-ended questions are also used to know farmers expectation from the extension service providers. A random sampling method was adopted to select 115 respondents. The samples were distributed all over the catchment area of which 90 were paid members of extension service providers and 25 were non-members. Data collected was analyzed using the Statistical Package for Social Science (SPSS).

RESULTS AND DISCUSSIONS

The respondents were interviewed to find the effectiveness of the paid advisory services and the data collected was tabulated. Table 1 revealed that 75.65% of the respondents have studied till primary or middle class. However, 15.65% of

Table 1: Characteristics of respondents (n=115 respondents)

Characteristics	Frequency	Percentage
Education level		
Illiterate	18	15.65
Primary	36	31.30
Middle	51	44.35
Matric & above	14	12.17
Land holding		
<2 ha	27	23.48
2-10 ha	83	72.17
>10 ha	5	4.35
Advisory services		
Paid members	90	78.26
Non-members	25	21.74

Table 2: Advisory services for crops (n=90 paid members)

Crops	Frequency	Percentage
Cotton	70	77.78
Sweet lime	9	10.00
Onion	7	7.78
Mango	1	1.11
Pome	3	3.33

Table 3: Critical needs of respondents (n=115 respondents)

Critical needs	Frequency	Percentage
Input supply linkage	42	46.67
Credit Purchase	34	37.78
Discount on fees	16	17.78
Financial services	23	25.56

the respondents were illiterate. The education level of the studied area was low. The data presented in table 1 revealed that 23.48% of the respondents had small landholding upto 2 ha. followed by 72.17% having landholding up to 2-10 ha. and large holders were 4.35% with land above 10 ha. Table 2 revealed that 77.78% of the paid members have taken advisory services for cotton crop which shows that cotton is the major crop in Aurangabad and main focus for extension service providers. The advisory services for other crops such as sweet lime, onion, pome and mango were taken by few respondents.

The perception of the farmers not subscribing to the advisory services revealed that the most frequently accessed source of 60% of them was 'other progressive farmers' and the 'dealer pro-

viding inputs'. The similar findings were reported by NSSO (2005, p.7). Of all the respondents 46.67% agreed that their critical need was supply of inputs followed by credit purchase for 37.78% of respondents which proves to be core constraints on which the advisory services provider should focus. We can say that majority of the respondents were aware of the crop advisory services but the fee charged and lack of financial services were additional constraints in approaching the crop advisory service providers (Table 3). Similar findings were confirmed by Gebremedhin, Hoekstra, and Tegegne (2006) who have indicated input supply linkage as critical need in case of Ethiopian farmers. Of all the respondents, 49.6% were aware about the advisory services mainly through fellow farmers and others were informed by advisory service providers or the leaflets. This shows that the dissemination channels for advisory services were not proper. The importance of ICTs was needed to be identified by service providers as an important means to increase agricultural productivity and consequently agricultural and rural income (Annamalai & Rao, 2003; Singh, 2006). Table 4 revealed that 23.3% respondents rated advisory services "good" and 51.1 % rated "fair" as it helped in reducing the cost of cultivation by utilization of better techniques. 11.1% rated these advisory services as "excellent" and 73.3% rated them as "good" due to regular follow up from the staff. The respondents were also satisfied with the recommendation and knowledge of the staff of the

Table 4: Distribution of respondents according to their perception about advisory services (n = 90 paid members)

	Excellent		Good		Average		Fair		Poor	
	f	%	f	%	f	%	f	%	f	%
Cost of cultivation and Market price	0	0.0	21	23.3	46	51.1	12	13.3	11	12.2
Reduction in Cost of cultivation	0	0.0	46	51.1	39	43.3	5	5.6	0	0.0
Market price of the produce										
Customer relationship status	6	6.7	70	77.8	14	15.6	0	0.0	0	0.0
Adviser's field visit	10	11.1	66	73.3	14	15.6	0	0.0	0	0.0
Follow-up by staff	16	17.8	56	62.2	18	20.0	0	0.0	0	0.0
Knowledge of staff										
Effectiveness of advice	9	10.0	47	52.2	34	37.8	0	0.0	0	0.0
Recommendation	6	6.7	55	61.1	29	32.2	0	0.0	0	0.0
Result after advice										

Table 5: Coefficient of correlation and significant levels

Variable 1	Variable 2	R
Recommendation	Increase in yield	0.767*
Result after advise	Increase in yield	0.879*

*Correlation is significant at the 0.01 level (2-tailed). n= 90 paid members

advisory service providers.

For describing relation between independent variables with dependent variable (increase in yield), Pearson's coefficient of correlation was used (Table 5). The correlation study shows that the recommendations by the advisory services and the results after advice have a positive correlation of 0.767 and 0.879 respectively with increase in yield showing the effectiveness of these crop advisory services. The farm demand for improved technology is significantly influenced by household utilization of advisory services and risks associated with farming. Results show that intensity of adoption first decreases, then increases with more advisory visits, suggesting that increased interaction of the farmer with service providers increases their awareness and knowledge regarding the use of improved technologies. Similar findings were reported by Isengildina, Pennings, Irwin, and Good (2004) and Mugisha & Diiro (2010). To describe the role of independent variable on dependent variable multivariate regression was used. The results of multivariate regression (Table 6 and 7) indicated that the cropping and harvesting method, credit access, input supply linkage, insurance, age, education and interaction with other farmers have the main role in showing the variations of attitude to adoption of the advisory services. It was concluded that different dissemination channels were not fully utilized in the area which hindered not only awareness level of the respondents but also adversely affected the rate of adoption level regarding the latest production technology.

CONCLUSIONS AND RECOMMENDATIONS

The advisory services were found to be a critical factor for increased adoption. It was concluded that the farmers who were visited regularly by crop advisory service providers, tended to adopt technology sooner and achieved

Table 6: Findings of multivariate regression

	Sum of Squares	ANOVAb Df	Mean Square	F	Sig.
Regression	3.808	9	.423	2.723	.007a
Residual	16.314	105	.155		
Total	20.122	114			

a. Predictors: (Constant), Input Supply Linkage, Education, Interaction with Farmers, Crop Pattern, Insurance, Harvesting Method, Age, Credit Access
b. Dependent Variable: Participation in Advisory Services

Table 7: Findings of multivariate regression

Independent variable	B	Std. Error	Beta	T	Sig
(Constant)	1.228	0.302		4.06	0.000
Age	0.004	0.006	0.067	0.716	0.476
Education	0.023	0.044	0.049	0.53	0.597
Insurance	0.047	0.11	0.039	0.424	0.673
Crop Pattern	-0.142	0.077	-0.164	-1.845	0.068
Harvesting Method	-0.128	0.08	-0.144	-1.609	0.111
Interaction with Farmers	-0.012	0.076	-0.014	-0.154	0.878
Credit Access	0.131	0.096	0.131	1.371	0.173
Input Supply Linkage	0.052	0.077	0.062	0.668	0.506

a. Predictors: (Constant), Input Supply Linkage, Education, Interaction with Farmers, Crop Pattern, Insurance, Harvesting Method, Age, Credit Access
b. Dependent Variable: Participation in Advisory Services

higher yields. Lack of knowledge was rarely cited as a reason for non-adoption of a specific technology. The most common reasons cited were economic considerations, climatic factors, unavailability of inputs, and lack of irrigation. This implies that rates of adoption were acceptable for those technologies that farmers easily perceived as advantageous; and lower for those that either were or seemed irrelevant to farmers' situations. The quality of the extension service should receive as much attention as the extension method. There was need to create awareness through various dissemination channels for faster adoption of technology. The critical need was input supply linkage and purchase on credit. The frequent extension staff visit and establishment of telecentres was required for faster dissemination of advisory services and latest technology with special emphasis on Integrated Water Management as the district has semi-arid type of climate.

REFERENCES

1- Alex, G., Zijp,W., & Byerlee, D. (2002). Rural extension and advisory services – new direction. (Rural strategy paper no. 9). World Bank, Washington DC.

2- Anandajayasekeram, P, Dijkman, J, Hoekstra, D, & Workneh, S. (2005, May 23-25). Past, present and future of extension services. Improving Productivity and Market Success (IPMS) of Ethiopian Farmers. Addis Ababa, Ethiopia: ILRI (International Livestock Research Institute).

3- Anderson, Jock R., & Feder, Gershon (2003, February). Rural extension services. (World Bank Policy Research Working Paper 2976).

4- Annamalai, K.,& Rao, S. (2003). What works: ITC's e-Choupal and profitable rural transformation: Web-based information and procurement tools for Indian farmers. World Resources Institute, Washington, D.C.

5- Axinn, G. H. (1988). Guide on alternative extension approaches. FAO (Food and Agriculture Organization of the United Nations), Rome, Italy.

6- Danida. (2002). Evaluation/impact study of four training projects for farm women in India (WYTEP, TANWA, TEWA, MAPWA). Copenhagen. Udenrigsministeriet.

7- Economic Survey. (2010-11).Chapter 8 Agriculture and Food Management, p. 188.

8- Evenson, Robert E., & Mwabu, Germano. (1998, September). The effects of agricultural extension on farm yields in Kenya. (Center discussion paper no. 798). Yale University, Economic Growth Center.

9- Farrington, J., Christoplos, L., Kidd, A., & Beckman, M. (2002). Extension, poverty, and vulnerability: the scope for policy reform: final report of a study for the Neuchatel initiative. (Working paper no. 155). Overseas Development Institute, London.

10- Feder, Gershon, Birner, Regina, & Anderson, Jock R. (2011). The private sector's role in agricultural extension systems: potential and limitations. Journal of Agribusiness in Developing and Emerging Economies. Retreived from: http://www.emeraldinsight.com/mobile/index.htm.

11- Gebremedhin, Berhanu, Hoekstra, D., & Tegegne, Azage. (2006). Commercialization of Ethiopian agriculture: Extension service from input supplier to knowledge broker and facilitator. IPMS (Improving Productivity and Market Success) of Ethiopian Farmers. (Project Working Paper 1). Nairobi, Kenya: ILRI (International Livestock Research Institute).

12- Goyal, A. (2010). Information, direct access to farmers, and rural market performance in central India. American Economic Journal: Applied Economics, 2 (3), 22-45.

13- Gustafson, D.J. (1994). Developing Sustainable Institutions: Lessons from Cross-Case Analysis of Agricultural Extension Programmes. Public Administration and Development 14(2), 121-34.

14- India, Planning Commission. (2005). Midterm appraisal of the 10th Five Year Plan (2002-2007). Retrieved from: http://planningcommission.nic.in/midterm/english-pdf/section -05.pdf.

15- Isengildina, Olga, Pennings, Joost M.E., Irwin, Scott H., & Good, Darrel L.(2004, May). Crop Farmers' Use of Market Advisory Services. AgMAS Project Research Report.

16- Mugisha, Johnny, & Diiro, Gracious. (2010). Explaining the Adoption of Improved Maize Varieties and its Effects on Yields among Smallholder Maize Farmers in Eastern and Central Uganda. Middle East Journal of Scientific Research, 5(1), 6-13.

17- Neuchatel Group. (2002). Common Framework on Financing Agricultural and Rural Extension,Swiss Centre for Agricultural Extension and Rural Development, Lindau. Retrieved from: http://www.g-fras.org/fileadmin/UserFiles/Documents/Frames-and-guidelines/Financing-RAS/Common-Framework-on-Financing-Extension.pdf

18- NSSO. (2005). Access to Modern Technology

for Farming, Retrieved from Press Information Bureau, Government of India, New Delhi website http://www.mospi.nic.in/nsso_press_note_22june05.htm.

19- Purcell, D.L., & Anderson, J.R. (1997). Agricultural research and extension: Achievements and problems in national systems. World Bank Operations Evaluation study, World Bank, Washington DC, USA.

20- Raabe, Katharina. (2008). Reforming the Agricultural Extension System in India : What Do We Know About What Works Where and Why? (IFPRI Discussion Paper 00775).

21- Rivera, W.M., & Zijp, W. (2002). Contracting for Agricultural Extension: International Case Studies and Emerging Practices. CABI publications: Wallingford.

22- Rivera, W.M., Qamar, K.M., & Crowder, L.V. (2001). Agricultural and Rural Extension Worldwide: Options for Institutional Reform in Developing Countries. Food and Agriculture Organization of the United Nations, Rome.

23- Schultz, Theodore W. (1975). The Value of the Ability to Deal with Disequilibria, Journal of Economic Literature, 12(3), 827-846.

24- Singh, P. (2006). Bengal Villages Get Connected, With a Little Help From an NGO . Retrieved from: http://www.indianexpress.com/full_story.php?content_id=85143.

25- Sulaiman V., Rasheed. (2003). Innovations in agricultural extension in India. FAO, Agriculture Extension and Research in India, IEG, World Bank Group

26- Sulaiman V., Rasheed. Extension Services in India: Emerging challenges and ways forward. Hyderabad: Centre for Research on Innovation and Science Policy (CRISP)(South Asia Rural Innovation Policy Studies Hub)

27- Sulaiman, V.R., & Hall, A.J. (2002). Beyond technology dissemination: reinventing agricultural extension, Outlook on Agriculture, 31 (4), 225-33.

28- Van der Ban, A.W., & Hawkins, H.S. (1996). Agricultural Extension (2nd edition). Ox

15

The Effect of Drought Stress on Germination and Early Growth of *Sesamum indicum* Seedling's Varieties under Laboratory Conditions

Mohammad Hossein Bijeh keshavarzi

Abstract

Environmental stresses specially drought, play an important role in decreasing plant growth, particularly during germination in dry and semi dry area. To considering the effect of drought stress caused by polyethylene glycol on germination and characteristics of 2 spices of Sesamum indicum, we had done factorial and complete accidental plot with 4 treatments and 3 times repetition. Experimental treatments included osmotic potential in 4 levels (0, -4, -6, -10 bar) which was produced by polyethylene glycol 6000 and 2 sesame species (Safi Abadi and Dezfol). All data had been analyzed by SAS software and comparison of means had been done by Duncan test at 5% probable level. The results showed that, percentage and speed of all spices' germination decline by osmotic potential enhancement. Other measured parameters such as radicle and coleoptile length, dry and wet weight declined by increasing osmotic potential as well.

Keywords:
Stress, Germination, Polyethylene glycol, Sesamum indicum

Young Researchers Club, North Tehran Branch, Islamic Azad University, Tehran, Iran.
* Corresponding author's email: keshavarzi64.mh@gmail.com

INTRODUCTION

Sesame (*Sesamum indicum*) is a flowering plant in the genus Sesamum. Numerous wild relatives of it occur in Africa and a smaller number in India. It is widely naturalized in tropical regions around the world and is cultivated for its edible seeds, which grow in pods.

Drought stress with osmotic materials for producing osmotic potential are one of the important studding approach about drought stress effects on germination. While sesame has a high resistance level in drought stress condition, germination and seedling stage make it more sensitive (Orruno and Morgan, 2007). These are 2 important and critical stages when crops are grown in arid zones. At this crucial and critical stages that ultimate yield of crops are determined (Hadas 1976).

Turk *et al.,* (2004) found that one of the reasons that can reduce or delay or even prevent germination is water stress. It also decreases germination rate and seedling growth rate. There were some studies that using local sesame from Nigeria which found that low level of drought stress hadn't any significant effect on germination, by increasing levels of drought germination and seedling growth reduced, on the other hand, drought stress level has negative correlation with germination and seedling growth (Heikal *et al.*, 1982; Mensah *et al.*, 2006).

Environmental stress during seed production period can has negative impact on next seeds quality (Sediyma *et al.*, 1972).

As polyethylene glycol can make semi natural environment has a wide range of application especially under laboratory conditions (Rade and Kar, 1995). As it has heavy molecular weight, can't transfer from cell wall. So it uses for adjusting water potential in germinating laboratories. Polyethylene glycol 6000 is more suitable than smaller molecules such as 4000 for making drought stress, because seed germination percentage in polyethylene glycol 6000 solvent is equal to the soil with the same water potential (Rade and Kar, 1995).

The purpose of this experiment is considering drought stress impacts which are results of polyethylene glycol on germination of 2 sesame species.

MATERIALS AND METHODS

This experiment was done in 2010 in agronomy laboratory of agriculture college, Zabol University. It was done in factorial and complete accidental plot with 4 treatments and 3 times repetition. Experimental treatments included osmotic potential in 4 levels (0, -4, -6, -10 bar) which was produced by polyethylene glycol 6000 and 2 sesame species (Safi Abadi and Dezfol).

Each experimental unit included 1 Petri dish with 8 cm diameters. For each treatment 25 of healthy sesame seeds from each species selected and was washed by hypochlorite sodium 10%, and was put on filter paper homogeneity then 5 ml of polyethylene glycol was added. Next Petri dishes doors were closed by Para film, and put in a small room. Germinated seeds were counted daily to determine germination race, and it took 6 days.

To determine needed poly ethylene glycol 6000 for solvent of each stress level, the following formula used separately: (Michel and Kaufman, 1973).

Water potential $= - (1.18 \times 10^{-2})$ C $- (1.18 \times 10^{-4})$ C2 $+ (2.67 \times 10^{-4})$ CT $+ (8.39 \times 10^{-7})$ TC2

C= polyethylene glycol concentration (gr/ water kg)
T= temperature (C) (Michel and Kaufmann, 1973)
In the following formula water potential unit is Mega Pascal.
0.1 Mpa = 1 bar or 105 pa = 1 bar

The following characteristics were studied:
Germination Percentage (GP):
From second day, the germinated seeds were counted daily in specific time. At that time, those seeds were considered germinated which their radical length was more than 3 mm.

Counting continued till we could count more germinated seeds and the resulted final counting considered as final germination percentage.
GP: Ni / N × 100
Ni: number of germinated seed till ith day)
N= total number of seeds.
Germination Race (GR):
In order that, from the second day to 6th once

a 24 hours we counted germinated seeds and its race was determined by Maguire equation (1962):

$$GR = \sum_{i=1}^{n} \frac{S_i}{D_i}$$

GR: Germination Race (number of germinated seed in each day)
Si: number of germination seeds in each numeration
Di: number of days till nth numeration.
n: number of numeration times.

At the end of experiment, 10 plants were selected from each Petri dish, radicle and pumule separated and measured. After this stage, each repetition put on filter paper separately and to dry it, we put it in oven and measure dry weight. SAS and MSTAT-C software was used to analysis.

RESULTS AND DISCUSSION
1. Germination Percentage (GP)
The effect of water potential was significant on germination percentage (P<001) (Table 1). Totally by increasing solvent concentration, ger-

mination percentage will decrease. Therefore, sesame species' germination percentages has significant differences in reaction to stress (P<0.01) (Table 1).

The interaction effects between two factors (water potential and species) were significant (P<0.01) on germination percentage (Table 1). Germination percentage was 100 in Dezfol specie and 0 water potential and it reached to zero in Safi Abadi specie and -10 water potential (Table 3).

2. Germination Race (GR)
The effect of water potential on germination race of sesame was significant also (P<0.01) (Table 1). Sesame species' germination races had significant differences in reaction to stress (P<0.01) (Table 1).

Interaction effects between water potential and sesame species were significant on germination race (P<0.01) (Table 1). Germination race in Dezfol and 0 water potential was 2.51, and it reached to 0 in Safi Abadi and -10 water

Table 1: Variance analysis of drought and variety effect on Sesamum indicum.

S.O.V	df	GP (%)	GR	RL (cm)	CL (cm)	FW (mg)	DW (mg)
Varieties	1	504.16**	1.79**	1.36**	1.66**	247.17**	1.042**
Water potential	3	10701.4**	5.16**	17.39**	3.87**	1366.03**	62.63**
Varieties × Water potential	3	262.5**	0.61**	0.24**	0.28**	40.84**	0.241**
Error	16	29.16	0.018	0.018	0.017	6.19	0.062
C.V (%)		10.53	12.85	8.12	15.69	17.58	9.35

Note: ** indicate significant difference at 1% probability level.
GR: Germination rate, GP: Germination percentage, CL: Coleoptile length, RL: Radicle length, FW: Fresh weight, DW: Dry weight.

Table 2: Mean comparison of seedlings germination and growth indexes under different drought levels and varieties of Sesamum indicum.

Treatment	GP (%)	GR	RL (cm)	CL (cm)	FW (mg)	DW (mg)
Varieties						
Dezfol	55.83a	1.32a	1.9a	1.11a	0.017a	0.0028a
Safi Abadi	46.66b	0.77b	1.41b	0.6b	0.011b	0.0024b
Drought level (Bar)						
0	94.16a	1.99a	3.85a	1.83a	0.034	0.007a
-2	76.66b	1.63b	2.1b	1.12b	0.015b	0.0032b
-4	33.33c	0.55c	0.6c	0.41c	0.006c	0.00037c
-8	0.83d	0.0045d	0.02d	0.014d	0.0001d	0.000005d

GR: Germination rate, GP: Germination percentage, CL: Coleoptile length, RL: Radicle length, FW: Fresh weight, DW: Dry weight.

Table 3: Mean comparison of interaction effects of drought stress and Sesamum indicum varieties in germination stage.

Varieties	Osmotic Potential (Bar)	GP (%)	GR	RL (cm)	PL (cm)	WW (mg)	DW (mg)
Dezfol	0	100a	2.51a	4.23a	2.267a	40a	7.3a
	-2	90b	2.22b	2.53c	1.583b	21c	3.7c
	-4	31.66d	0.54e	0.76e	0.566e	8.23e	0.456e
	-8	1.67e	0.009f	0.4g	0.028g	0.206g	0.01f
Safi Abadi	0	88.33b	1.48c	3.46b	1.407c	29.33b	6.66b
	-2	63.33c	1.05d	1.66d	0.66d	10.2d	2.83d
	-4	35d	0.56e	0.53f	0.266f	4.23f	0.3e
	-8	0e	0f	0g	0g	0g	0f

GR: Germination rate, GP: Germination percentage, PL: Coleoptile length, RL: Radicle length, WW: Wet weight, DW: Dry weight.

potential (Table 3).

One of the important indexes for assessing drought stress tolerance is germination race; species with high germination race are likely to become green rapidly than other spices under drought stress. It seems that high speed in water absorbance leads to increase in germination race (Marchner, 1995). If there appear any interfere in water absorbance or it happiness slowly, germination metabolically activities will do slowly in seeds, so germination race will decrease (Abnos, 2001).

3. Radicle and coleoptile length

Water potential had significant difference on radicle and coleoptile length (P<0.01) (Table 1). By increasing solvent concentration, radicle and coleoptile length decreased.

Interaction effects between water potential and spices had significant difference on radicle and coleoptile length of sesame seedlings (P<0.01) (Table 1).

4. Dry and FrEesh weight of seedlings

Means comparison showed that among spices, the driest weights were related to Dezfol and Safi Abadi orderly (Table 2).

Interaction effects comparison showed that Dezfol spice and 0 level of stress had the most dry and fresh weight of seedling, and Safi Abadi spice and -10 level of stress had the least dry and fresh seedlings' weight (Table 3).

Some researchers reported that under stress conditions, coleoptile growth was more than

radicle growth and weight decrease mote than length. But others believe that stress will decrease radicle length more than coleoptile, but weight remain stable (Sediyma et al., 1972).

Reduction in the yield percentage of these initial processes can be referred to lower in fusibility in water uptake of seed under stress condition (Turk et al., 2004). Germination process consists of 2 stages, firstly enzymatic hydrolysis of stored material and secondary building of new tissue by hydrolysis (Bahrami et al., 2012). Moisture deficit can affect enzymatic activity and consequently germination percentage decreases under the more negative osmotic potential.

Water absorption and seed swelling considered as the first step in germination, while cell division and elongation are last steps. By decreasing osmotic potential, water absorption will decrease as well as cell division (Zaefizadeh et al., 2011).

Another important feature for drought stress is root length, because it has direct contact with soil and water absorbance. As result, root length make important clue to a plants response to drought stress (Mostafavi et al., 2011).

The primary organ of young plant extension is hypocotyls, which after emergence of radicle emerge; in drought stress, radicle will develop faster than hypocotyls to compensate for water stress (Zhu et al., 2005). So at germination stage growth of hypocotyls and radicle reflects tolerance of shoot to drought (Shi and Ding 2000). From aforementioned studies can conclude that reduction of root and shoot could be referred to reduction of cell division which is caused by water stress. In addition, cause of root and shoot dry weight's reduction is reductions in root and shoot lengths.

REFERENCES

1- Abnos, M, (2001). Study of physiological effects of drought stress on germination and seedlings of lentil spices, Msc dissertation of plants physiology, Science College, Ferdosi University, Mashhad. 147 Pp.

2- Bahrami, H, J, Razmjoo, and A, Ostadi Jafari, (2012). Effect of drought stress on germination and seedling growth of sesame cultivars (*Sesamum indicum* L.). International Journal of AgriScience 2(5): 423-428.

3- Hadas, A, (1976). Water uptake and germination of leguminous seeds in soils of changing matrix and osmotic water potential. J Exp Bot. 28: 977-985.

4- Heikal, M.M, M.A, Shaddad, and A.M, Ahmed, (1982). Effect of water stress and gibberellic acid on germination of flax, sesame and onion seed. Biol Plant 24:124-129.

5- Marchner, H, (1995). Mineral Nutrition of Higher Plants. Academic Press.

6- Mensah, J.K, B.O, Obadoni, P.G, Eroutor, and F, Onome-Irieguna, (2006). Simulated flooding and drought effects on germination, growth, and yield parameters of sesame (*Sesamum indicum* L.). Afr J Biotechnol. 5(13): 1249-1253.

7- Michel, B.E, and M.R, Kaufmann, (1973). The osmotic potential of Polyethylene glycol 6000. Plant physiology 51: 914-916.

8- Mostafavi, K, H, Sadeghi Give, M, Dadresan and M, Zarabi, (2011). Effects of drought stress on germination indices of corn hybrids (Zea mays L.) Int J AgriSci. 1(2): 10-18.

9- Orruno, E, and M.R.A, Morgan, (2007). Purification and characterization of the 7S globulin storage protein from sesame (Sesamum indicun L.). Food Chem. 100: 926-934.

10- Rade, D, And R.K. Kar, (1995). Seed germination and seedling growth of mange bean under water stress induced by PEG 6000. Seed Sci and Tecnol . 23: 301-308.

11- Sediyma, T., A.A, Cardoso, and C. Vieira, (1972). Tests Preliminares sobre os efeitos do ratardamento da colheita da soja cultivar vicoja. Rev. Ceres 19: 306-310.

12- Shi, J.Y, and G.J, Ding, (2000). Effects of water stresses on Pinus massoniana seeds from different provenances. J Mountain Agr Bio. 19: 332-337.

13- Turk, M.A, A, Rahmsn, and M Lee K.D, Tawaha, (2004). Seed germination and seedling growth of three lentil cultivars under moisture stress. Asian J Plant Sci. 3: 394-397.

14- Zaefizadeh, M, S, Jamaati-e-Somarin, R, Zabi-hi-e-Mahmoodabad, and M, Khayatnezhad, (2011). Discriminate analysis of the osmotic stress tolerance of different sub-cultivars of durum wheat during germination. Adv Environ Biolo. 5(1): 74-80.

15- Zhu, J.J, H.Z, Kang, H, Tan, M.L, Xu, and J, Wang, (2005). Natural regeneration characteristics of natural Pinus Sylvestris var. Mongolia forests on sundy land in Honghuaerji. China J For Res. 16: 253-259.

16

Employee Job Autonomy and Control in a Restructured Extension Organization

Mary S. Holz-Clause[1], Vikram Swaroop Chandra Koundinya [2], Nancy K. Franz [3] and Timothy O. Borich [4]

Abstract

*Keywords:
Extension, restructur-
ing, autonomy, control,
programming*

This descriptive cross sectional census study identified the perceptions of Extension and Outreach employees of Iowa State University in the United States about job autonomy and control after two years of a major restructuring. Employees perceived autonomy and control over expressing views and ideas about their work and spending time on the job but perceived little influence over budget allocations and shaping organizational strategies. They felt administrators and external funding sources influenced programming. They perceived contributing most to program implementation and marketing. The findings from this study have implications for operations and programming in Extension and other organizational settings.

[1] *Vice President for Economic Development, University of Connecticut, 304 Gulley Hall, Storrs, CT 06268, U.S.*

[2] *Postdoctoral Research Associate, ISU Extension and Outreach, 220 Curtiss Hall, ISU, Ames, IA 50011, U.S.*

[3] *Program Director for Families, ISU Extension and Outreach, 111 MacKay, Ames, IA 50011, U.S.*

[4] *Program Director for Communities, ISU Extension and Outreach, 126 Design, Ames, IA 50011, U.S.*

* *Corresponding author's email: mary.holz-clause@uconn.edu*

INTRODUCTION

The United States economic recession that began in 2007 and the subsequent financial crisis of 2008 left many public organizations reassessing their financial foundations and value proposition. In response, these organizations employed various strategies to maintain their services (Holz-Clause *et al.*, 2012). Organizational restructuring, one strategy adopted during difficult financial times, has become an important strategic response to budget cuts (McKinley & Scherer, 2000). This strategy has been utilized in the Cooperative Extension System (CES) within the United States, as state and federal budgets have been declining in relative terms. Many CESs have restructured their services in the past 20 years with varying degrees of success (Ahmed and Morse, 2010; Bartholomew and Smith, 1990; Hutchins, 1992; Jayaratne and Gamon, 1998; Rockwell *et al.,* 1993; Schafer, 2006; Schmitt and Bartholomay, 2009; Suvedi *et al.*, 2000; Tondl, 1991).

A recent example of restructuring is a CES at Iowa State University. Financial realities required a plan that addressed revenue reductions while maintaining an orderly reduction to staff and consistent delivery of programs. The goal was a more efficient administrative structure. The argument was presented that it was appropriate to move from the anachronistic geographically focused structure to an issues-based one. The result was a regional administrative model with far more local/county control and responsibility. The main components of the regional model included (1) elimination of the five area administrative positions and associated office and staffs, (2) elimination of all 100 county extension education director positions, (3) creation of 20 new Extension regions with 20 regional directors (REEDs) overseeing the operations and programming of group of counties in the region, and (4) reduction of the five main Extension programs to three by combining 4-H Youth Development with the Families program area and Community Economic Development program area with the Business & Industry program area.

This restructuring resulted in layoffs while concurrently changing work jurisdictions and new partnerships for program implementation. This meant a major realignment of employees and a disorienting effect on the organization. McKinley and Scherer (2000) stated that organizational restructuring of any kind results in both anticipated and unanticipated outcomes, and organizations should consider the effects of restructuring on job performance of its employees (Jayaratne & Gamon, 1998). It is only through subsequent assessment that we understand the accuracy of what was anticipated or the nuance of what was not.

Research on impact of organizational restructuring on employees shows both positive and negative outcomes, but organizational restructuring by itself is not good or bad (McKinley and Scherer, 2000). Schmitt and Bartholomay (2009) found that regionalization of Extension resulted in improved work attributes of Minnesota Extension employees. Similar results were recorded by Ahmed and Morse (2010). However, Jayaratne and Gamon (1998) found that restructuring Extension in Illinois resulted in increased anxiety levels in employees. Similarly, McKinley and Scherer (2000) stated that organizational restructuring may lead to a sense of disorder in the organization and a bifurcation between managers and other employees. These results tend to only inform us that the outcomes are uncertain since the underlying circumstances of leadership, staff demographics, client perceptions, etc. are among the many variables.

Of the various factors that contribute to organizational success, job autonomy and control are important for sustaining and improving employee contribution to the organization. Kroth and Puets (2011) stated that job autonomy is one of the required factors for creating a supportive work environment. Similarly, Extension educators identified lack of job autonomy and control as a major challenge to their work (Kuetelik *et al.*, 2002). Schmitt and Bartholomay (2009) found that regional educators perceived a significant gain in their autonomy whereas local educators perceived no such difference as a result of the regionalization of Extension services. This study identified the perceptions of Iowa State University Extension and Outreach employees about three aspects related to their job autonomy and control after working in a restructured regional model for two years. The purpose was to create a baseline that can be used in the future to gauge employee perceptions about job autonomy and control.

Objectives of the Study

The study had three specific objectives:

1- Identify the perceptions of employees about their job autonomy and control.

2- Identify the perceptions of employees about the influence different entities have on Extension and Outreach programming.

3- Identify the perceptions of employees regarding their contribution to organizational success or decline.

MATERIALS AND METHODS

A descriptive cross-sectional census survey was used for this study. The population consisted of all 956 paid employees working for Iowa State University Extension and Outreach, which included county-based employees and university paid faculty and staff both located on campus and in the county offices. The Institutional Review Board at the Iowa State University approved this study. An electronic questionnaire developed using SurveyMonkey® was employed for this study. The questionnaire was developed by the researchers and validated for face and content validity by an expert panel consisting of select leadership team members of the Iowa State University Extension and Outreach.

The questionnaire was pilot-tested with randomly selected employees, and the data were used to establish the reliability of the questionnaire. Cronbach's α was computed from the pilot test data and values of 0.929, 0.961, and 0.962 were reported for the three sections, respectively indicating 'excellent' reliability (George & Mallery, 2003). The participants were emailed the questionnaire, including an introductory message informing them of the purpose of the study. This introductory email indicated that their participation in the study was voluntary and that they could withdraw at any time. Par-

ticipants' consent for the study was assumed if they filled out and returned the questionnaires. After that, a total of three follow-ups (Dillman, 2007) were sent weekly.

A four point Likert-type scale was used for all three sections of the survey. There were 8 (Section 1: perceptions about job autonomy and control), 14 (Section 2: perceptions about influence of different entities on programming), and 14 (Section 3: perceptions about their contribution to organizational success or decline) items under each section, respectively. For measuring the perceptions about job autonomy and control and the perceived contribution to organizational success or decline the scale used was from 1 (None) to 4 (Significant); 1 (No influence) to 4 (Significant influence) was used for measuring the perceived influence of different entities on programming. A four-point scale was employed so employees take either a positive or a negative stance, and not stay undecided about any statement in the questionnaire, as this study was conducted to create a baseline that can be used in the future to gauge employee perceptions about job autonomy and control.

Data were analyzed using PASW® Statistics 18. Descriptive and inferential statistics were used in the data analysis. Frequencies (f), mean (M), standard deviation (SD), and percentages (%) were used for analyzing the perceptions and demographic information of the participants. An independent samples t-test was used to test for any statistically significant differences between early and late respondents. Early respondents were operationally defined as those participants who responded to the first mailing and the first follow-up, and those who responded after that were considered as late respondents.

RESULTS

Four hundred fifty-four employees responded to the survey for a response rate of 47.5%. A majority were female (70.3%). Forty-one percent of the employees were based in counties followed by 32.2% on campus and 26.8% in field offices. The employees had a

Table 1: Perceptions of Employees Regarding Their Job Autonomy and Control

Statement related to job autonomy and control	f				M	SD	N
	1	2	3	4			
To what degree do you comfortable expressing ideas and views about your work	6	66	234	141	3.14	0.70	447
To what degree you feel you have control over the time you spend on your job	3	86	215	140	3.10	0.72	444
To what degree do you have control to make decisions about your work	2	70	269	104	3.06	0.63	445
To what degree do you feel you can shape the organization's programming strategies	56	191	162	34	2.89	0.80	443
To what degree do you feel you contribute to Extension's fiscal health	40	142	198	61	2.63	0.83	441
To what degree do you feel you have control of the use of the money you raise	79	137	155	58	2.44	0.94	429
To what degree do you feel you can shape the organization's operational strategies	90	232	105	19	2.11	0.77	446
To what degree do you feel you have control over Extension budget allocations	217	182	32	11	1.63	0.72	442

Note. 1 = None, 2 = Little, 3 = Good, 4 = Significant

wide range of work experience with 29.6% more than 20 years followed by 21% with 3-5 years, 16.7% with 1-2 years, 16.7% with 6-10 years and 16% with 11-20 years of work experience.

Objective 1: Identify the Perceptions of Employees about their Job Autonomy and Control

Employees clearly articulated three areas of their work where they perceive significant autonomy and control: 1) expressing ideas and views about their work, 2) spending time on the job, and 3) making decisions about their work. They also indicated that they felt little autonomy or control over: 1) budget allocations and 2) shaping the organization's operational strategies (Table 1).

Objective 2: Identify the Perceptions of Employees about the Influence Different Entities Have On Extension and Outreach Programming

Employees perceived program directors to be

Table 2: Perceptions of Employees about the Influence of Different Entities on Extension and Outreach Programming

Entities	f				M	SD	N
	1	2	3	4			
Program Directors	13	58	216	137	3.12	0.75	424
External Funding Sources	10	93	185	140	3.06	0.79	428
Campus faculty/staff	17	97	191	126	2.98	0.82	431
Field Specialists	8	89	245	89	2.96	0.69	431
Vice President of Extension and Outreach	26	100	159	134	2.95	0.89	419
Clients	7	123	236	66	2.83	0.69	432
Key Constituency Groups	17	112	209	74	2.82	0.76	412
Programmatic Partners	10	112	235	56	2.81	0.68	413
County Staff	19	146	199	63	2.71	0.76	427
County Extension Councils	11	171	188	60	2.69	0.73	430
USDA	30	149	157	72	2.66	0.85	408
Regional Extension Education Directors (REED)	32	146	192	52	2.62	0.79	422
ISU Provost	46	158	124	85	2.60	0.93	413
ISU President	45	165	122	82	2.58	0.92	414

Note. 1 = No influence, 2 = Little influence, 3 = Good influence, 4 = Significant influence

Table 3: Perceptions of Employees Regarding Their Contribution to Organizational Success or Decline

Item	f				M	SD	N
	1	**2**	**3**	**4**			
Program implementation	30	72	181	120	2.97	0.88	403
Marketing programs	23	90	186	105	2.92	0.84	404
Marketing the organization	15	96	198	96	2.92	0.78	405
Network development	26	100	183	91	2.84	0.84	400
Participating in professional development	22	95	218	71	2.83	0.77	406
Partnership development	36	83	198	87	2.83	0.86	404
Program development	45	80	191	86	2.79	0.90	402
Sharing program impacts	37	108	196	59	2.69	0.83	400
Evaluating programming	38	117	190	61	2.67	0.84	406
Using program delivery innovations	32	114	190	53	2.67	0.81	389
Conducting program needs assessments	46	116	185	58	2.62	0.86	405
Using program content innovations	37	126	182	46	2.60	0.81	391
Obtaining contracts, grants, fees, gifts	61	122	148	75	2.58	0.95	406
Personnel recruitment	78	151	126	47	2.35	0.92	402

Note. 1 = None, 2 = Little, 3 = Good, 4 = Significant

the most influential on Extension and Outreach programming followed by external funding sources. They perceived university administrators other than Vice President of Extension and Outreach to be the least influential entities in programming (Table 2).

Objective 3: Identify the Perceptions of Employees regarding Their Contribution to Organizational Success or Decline

Employees perceived themselves contributing positively to program implementation followed by marketing programs and the organization. On the contrary, employees felt they contributed the least towards personnel recruitment and in obtaining contracts, grants, fees and gifts for the organization (Table 3).

Statistically significant differences existed between early and late respondents for the variables 'County Extension Councils,' 'Regional Extension Education Directors' and 'Vice President of Extension and Outreach' under the entities influencing programming, and "using program content innovations' under self-contribution to organizational success or decline at 0.05 level of significance. Late respondents recorded higher mean perception scores than the early respondents on these four variables. The findings were not generalized to the total population on these four variables.

CONCLUSIONS AND RECOMMENDATIONS

Three conclusions were drawn based on the findings from each of the three objectives of this study. First, the employees feel autonomy and control over their programming but when it comes to influencing budget and organizational strategies, they do not feel empowered. Administration should consider engaging employees more in these organizational matters.

Secondly, employees feel program directors and external funding sources to be the major entities influencing Extension and Outreach programming. Further, employees do not yet understand the role of the 20 regional directors (REEDs) in the newly restructured organization. The REED's position has been evolving and changing the past two years, as the members of the organization come to define their roles and responsibilities, including elected county extension council members and county-paid staff. The REEDs continue to define their role and job descriptions. REEDs should communicate with constituents and colleagues about what they do to ensure they are meeting the organizational and client needs.

Third, employees were contributing more to program implementation and marketing compared to other extension educational processes. They were not contributing on the same level to processes such as evaluating programs and conducting needs assessments which are critical for continued programs implementation and continuation. Reasons for this need to be explored and addressed.

Overall, the findings revealed that employees don't feel equally empowered and contribute the same to all aspects related to Extension operations and programming. Extension administrators should consider these factors while making strategic decisions, and design programs accordingly. The findings may also have implications for designing professional development programs for employees. Many public Extension systems worldwide are facing competition from private Extension providers (Bennett, 1996) and reduced public funding (Bennett, 1996; Evans-Brown, 2012; Schindler, 2011; South Dakota State University Extension, 2011). The results from this study and the regional model may have implications for such countries. The results are equally applicable to any organizational setting outside of Extension.

REFERENCES

1- Ahmed, A., & Morse, G. W. (2010). Opportunities and threats created by extension field staff specialization. Journal of Extension [On-line], 48(1) Article 1RIB3. Available at: http://www.joe.org/joe/2010february/rb3.php

2- Bartholomew, H. M., & Smith, K. L. (1990). Stresses of multicounty agent positions. Journal of Extension [On-line], 28(4) Article 4FEA2. Available at: http://www.joe.org/joe/1990winter/a2.php

3- Bennett, C. F. (1996). Rationale for public funding of agricultural extension programs. Journal of Agricultural and Food Information, 3(4), 3-25.

4- Dillman, D. A. (2007). Mail and internet surveys. The tailored design method (2nd ed.). Hoboken, NJ: John Wiley & Sons, Inc.

5- Evans-Brown, S. (2012, April 5). UNH Cooperative Extension finds efficiencies, cuts service. New Hampshire News. Retrieved from http://www.nhpr.org/post/unh-cooperative-extension-finds-efficiencies-cuts-service

6- George, D., & Mallery, P. (2003). SPSS for windows step by step. A simple guide and reference. 11.0 update (4th ed). New York, NY: Pearson Education, Inc.

7- Holz-Clause, M. S., Koundinya, V.S.C., Glenn, S., & Payne, J. M. (2012). Regionalization of the Iowa State University Extension system: Lessons learned by key administrators. International Journal of Agricultural Management and Development, 2(1), 33-40.

8- Hutchins, G. K. (1992). Evaluating county clustering. Journal of Extension [On-line], 30(1) Article 1FEA5. Available at: http://www.joe. org/joe/ 992spring/a5.php

9- Jayaratne, K. S. U., & Gamon, J. (1998). Effects of restructuring on the job performance of extension educators: Implications for in-service training. Journal of Agricultural Education, 39(4), 45-52. doi: 10.5032/jae.1998.04045

10- Kroth, M., & Puets, J. (2011). Workplace issues in Extension—A Delphi study of extension educators. Journal of Extension [On-line], 49(1) Article 1RIB1. Available at: http: //www. joe.org / joe/2011february/rb1.php

11- Kutelik, L, M., Conklin, N, L., & Gunderson, G. (2002). Investing in the future: Addressing work/life issues of employees. Journal of Extension [On-line], 40(1) Article 1FEA6. Available at: http://www.joe.org/joe/2002february/a6.php

12- McKinley, W. & Scherer, A. G. (2000). Some unanticipated consequences of organizational restructuring. Academic Management Review, 25(4), 735-752.

13- Rockwell, S. K., Furgason, J., Jacobson, C., Schmidt, D., & Tooker, L. (1993). From single to multicounty programming units. Journal of Extension [On-line], 31(3) Article 3FEA4. Available at: http://www.joe.org/joe/1993fall/a4.php

14- Schafer, S. R. (2006). Clientele perceptions of the University of the Wyoming Cooperative Extension Service livestock program. Journal of Extension [On-line], 44(2) Article 2RIB6. Available at: http://www.joe.org/joe/2006april/rb6.php

15- Schindler, M. (2011, February 23). Extension programs face cuts. The Cornell Sun. Retrieved from http://cornellsun.com/node/46002

16- Schmitt, M. A., & Bartholomay, T. (2009). Organizational restructuring and its effect on agricultural extension educator satisfaction and effectiveness. Journal of Extension [On-line], 47(2) Article 2RIB1. Available at: http://www.joe.org/joe/2009april/rb1.php

17- South Dakota State University Extension (2011). Stewards of progress. College of Agricultural and

Life Sciences. Retrieved from http://www. sdstate. edu/abs/iGrow/upload/Stewards-of-Progress.pdf

18- Suvedi, M., Lapinski, M. K., & Campo, S. (2000). Farmers' perspectives of Michigan State University Extension: Trends and lessons from 1996 and 1999. Journal of Extension [On-line], 38(1) Article 1FEA4. Available at: http://www. joe.org/joe/2000february/a4.php

19- Tondl, R. M. (1991). Climate for change in Extension. Journal of Extension [On-line], 29(3) Article 3FEA4. Available at: http://www.joe.org/joe/1991fall/a4.php

17

Estimation of Potential Evapotranspiration and Crop Coefficient of Maize at Rupandehi District of Nepal

Govinda Bhandari

Keywords:
Maize, Lysimeter, Potential evapotranspiration, Crop coefficient, Aridity index, Nepal

Abstract

This study was conducted to determine the potential evapotranspiration (PET) of maize, the crop coefficient (Kc) under full water requirement as well as the cause of decrease in maize yield. It was determined that the seasonal PET of maize is about 486.6 mm. The Kc under full water supply was found to be: 0.11, 0.35, 1.51 and 0.34 for initial, development, mid-season and the late season stages respectively. The study also revealed that maintenance of sufficient moisture need of maize has a significant effect on growth, development and fruiting of the crop.

Environment Professionals' Training and Research Institute Pvt. Ltd. Kathmandu, Nepal
* Corresponding author's email: govinda.eptri@gmail.com

INTRODUCTION

Potential evapotranspiration (PET) is an important factor in the estimation of water requirement of crops. In the FAO Irrigation and Drainage paper No. 24 on Crop Water Requirements, four methods were given by Doorenobs and Pruit (1977) for the calculation of PET ; Blaney-Criddle, radiation, Penman (adjusted) and pan evaporation. Maize (*Zea mays* L.) is the second important crop in Nepal after rice in terms of area and production. It is grown in the sub tropical to cool temperate climates. For higher yields, crop water requirement is 500-600 mm depending upon the climate and duration of the crop, there should be adequate water during the crop establishment period. Water deficit during the grain filling period results in reduced grain weight. However, during the maturity and harvesting period, rainfall has negative impact on maintaining grain quality (Nayava and Gurung, 2010). Maize is grown during April- July in Terai. During this period, about 500 to 800 mm rainfall (R) occur in the Terai. Annual mean temperature (T) during the growing period of maize in Terai is 21^0C to 33^0C. In fact Terai region has only 9.1% of the total maize area and contributes 20.4% of the total maize production in the country. During the last 36 years period from 1970/71 to 2007/08, the production of maize in the Terai region increased from 230,700 MT to 383,141 MT. It is interesting to note that the maize yield in Terai was initially low. In Terai, area under maize increased by 60% and its yield increased only 71%. The average yield of maize from 1970/71 to 2007/08 in Rupandehi district is 1822 kg ha^{-1}. The yield of maize in Rupandehi district is observed lowest in 1996 and the highest in 1976 which is 934 kg ha^{-1} and 3241 kg ha^{-1} respectively (MoAC, 2007). Maize is planted in all three seasons (summer, winter and spring) in the Terai region. The planting of summer maize starts from April to May and harvests on August to September. The growing season of winter maize covers from November / December to March/ April. Similarly, the growing season of spring maize is in March/April and harvests on June to early July. For the summer and winter crops, farmers prefer to grow long duration varieties, where as for the spring maize short duration varieties are preferred. Winter and spring maize is planted, wherever the irrigation facilities are available. After the introduction of high yielding varieties, the yield increased in Terai comparatively more than those of the hills and mountain (Nayava and Gurung, 2010).

The maintenance of low plant population in maize fields is one of the major factors affecting low productivity of maize in Nepal. Since yield of a crop is the result of final plant population, the establishment of optimum plant population is essential to get maximum yield. The competition between plants may not occur, and resources are not efficiently used at very low plant population. Under low plant population, grain yield is limited by inadequate number of plants whereas at higher plant population, it declines mostly because of an increment in the number of aborted kernels and/or barren stalks. At higher plant population, the rate of yield reduction is in response to decreasing light, moisture nutrient and other environmental resources available to each plant. Nepalese farmers have been practicing broadcasting method of seed sowing, which can not maintain uniform plant population in the field. The maintenance of plant population of maize in farmer's field condition is lower than the recommended level; and hence the yield is reduced. This is one of the causes of lower maize yield in Nepal (Shrestha and Timsina, 2011).

Recently the effect of climate change to agriculture in Nepal was studied by different researchers. The Asia Pacific Network report shows, when the temperature rises up to 1^0C, there is a positive role in percentage change in maize yield in all the agro-ecological zones. When the temperature rises up to 2^0C and at the same time carbon dioxide is doubled, the yield will decrease in Terai, hills will not be much affected. On the contrary, mountain environment will have better yield if the carbon dioxide is doubled and temperature rises up to 4^0C, the maize yield in Terai will suffer 25% lesser than the present yield (Nayava and Gurung, 2010). Overall rainfall during summer monsoon of the year 2005-06 and 2006-07 was about 16% below the normal which reduced cultivation area of agricultural production in the country (MoE, 2011).

Scientific statements regarding changing climate of Nepal are pronouncedly focused on temperature rise at 0.06 °C per annum. Such a rise in average temperature is variable across the country (Gautam and Pokhrel 2010), higher in the mountains and Himalaya (0.08°C) as compared to low-lying Terai (0.04°C). As a result, series of speculation have been reported such as days and nights becoming warmer, retreating snow lines, increased evapotranspiration and peak runoff, decreasing regular water discharge in streams and recharge of natural water stores, changing water tables and changing pattern of precipitation in terms of form, season, duration and amount (MoE, 2010 a; Gautam and Pokhrel, 2010). Also reported are decreasing total rainy days and increasing number of drier days (evapotranspiration>precipitation) and days receiving over 100 mm rain (MoE, 2010 a; Practical Action, 2009). Precipitation, though not much varying in total amount, is mentioned being erratic and ill timed (Gurung, 2008). Weather variability associated with rising temperature and changing pattern of precipitation is expected to have utmost adverse impacts on various components of agricultural systems. The impacts, though expected to become higher in the mountains compared to low lying terai region, are detrimental to both the regions and ultimately to agricultural production, food security and people's economic sustenance.

The aim of this study is to estimate Kc; Aridity index, AI; and PET of maize using a Lysimeter and Blaney-Criddle method that could be used in Nepal.

MATERIALS AND METHODS
General description of the study area and a lysimeter.

The study area was located at National Wheat Research Program (NWRP), Bhairahawa in Rupandehi district as shown in Figure 1. It is situated between 27° 32' N latitude and 83° 25' E longitude. It is 105 m above the sea level.

In the terai, mostly sandy loam and loam types of soils are reported. Farmers reported black clay as the most fertile soil. However, this soil is not suitable for maize cultivation except in drier years, as it holds water longer than other soil types and tends to waterlog. Land preparation is difficult in clay soils, especially red clay. Brown and gray colored loams are the most suitable for maize cultivation (Paudyal et al., 2001). Loam and clay loam soil textures are the dominant soils of the study area. Loamy soil is easy to plow but yield declines substantially if rainfall is low.

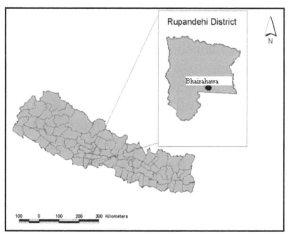

Figure 1: Location map of the study area in a Rupandehi district, Nepal

Experimental design and field layout

A lysimeter is made from readily available materials, plastic bucket with an area of 5811 square centimeter and 10 cm deep and is kept at one corner of the maize growing plot. The first bucket (B1) with tiny holes is completely filled with soil and leveled to the ground surface and holds maize seeds and the second bucket (B2) underground with a hole is connected with a pipe to pass out the infiltrated water to the receiving vessel which is kept deep underground as shown in figure a and b respectively. The receiver vessel kept underground and the B2 are connected with a water pipe so as to collect the percolated water. (Figure a and b)

Figure a: Water pipe connecting to B2 bucket

Figure b: A completely installed lysimeter

Figure b: B1 bucket with sowed maize seeds and a receiver vessel over B2 bucket which is connected with a receiver vessel

Growth stages

Four growth stages are considered. They are initial stage, developmental stage, mid-season stage, and late season stage. The initial stage lasted for 10 days (May 22 – May 31, 2011). The developmental growth stage lasted for 30 days (June 1– June 30, 2011). The mid-season growth stage (flowering and fruiting) stage lasted for 31 days (July 1 – July 31, 2011) and the late season stage lasted for 14 days (August 1– August 14, 2011). (Figure a and b)

Planting

Rampur Composite, a major downy mildew disease resistance variety of maize is sown on 22 May, 2011 as in fig a. Screening of this variety against downy mildew disease was carried out at Suwan farm, Thailand in 1981/82 (Rajbhandari, 1982). (Figure a and b)

Figure a: Mid-season growth stage

Figure a: Sowed maize seeds

Figure b: Late season stage

Irrigation regime

One liter of water is maintained daily except the rainy days so as to provide the sufficient moisture for the growth of maize in a lysimeter. Irrigation days amounted to 42 days and rainy days amounted to 43 days out of the 85 days of the total growing period. The percolated water is used for the irrigation to conserve the chemical composition of the soil.

Potential evapotranspiration

Evapotranspiration is one of the major contributors of the total irrigation water requirement. The amount of evapotranspiration is largely governed by weather parameters. Evapotranspiration might be potential or actual. Potential evapotranspiration assumes the continuous supply and no deficit of moisture in the air (WMO, 1994).

The measurement of PET includes the moisture evaporated to the atmosphere from plants and soil. As the soil and vegetation is confined within a small tank (the lysimeter) the measurements are made of the water input: Rainfall (R) and Additional water (A) and output: Percolated water (P) collected in the receiving vessel, PET can be estimated from the equation below:

$$PET = R + A - P \dots\dots\dots\dots\dots\dots\dots\dots (1)$$

Crop coefficient

K_c is calculated by using the value of PET (as obtained in equation 1) in a Blaney-Criddle formula (as in equation 2)

$$PET = 2.54 \, Kc \, F \dots\dots\dots\dots\dots\dots\dots (2)$$

and $F = \sum Ph \, Tf / 100$,

where,

K_c = an empirical coefficient, depends on the type of the crop

P_h = monthly percent of annual day time hours, depends on the latitude of the place

T_f = mean monthly temperature in °F

F = sum of monthly consumptive use factors for the period

Aridity Index

AI represents the severity of dryness of a region. Aridity is defined as the more or less repetitive climatic condition, which is characterized by a lack of water (Perry, 1986). It should be noted that aridity can be considered on seasonal or monthly basis (Coughlan, 2003). The AI ranges from 0.05 to 0.65 for the dry seasons. AI less than 0.65 correspond to Dry lands that, according to the United Nations Convention to Combat Desertification (UNCCD), may suffer desertification processes. So, AI should be greater than 0.65 during the maize growing seasons. In this study, UNEP aridity index (Hare, 1993) is used to estimate the AI which can be expressed as:

$$AI = P/PET \dots\dots\dots\dots\dots\dots\dots\dots\dots (3)$$

where, P=Precipitation in mm and PET=Potential Evapotranspiration in mm.

Other data collected

Five plants are selected randomly from the field and compared with the five plants of the lysimeter to compare the different parameters.

Height and width: Plant height and width of stalk is measured by a measuring tape.

Corn yield: Late April planted corn yields the highest and produces the most profit per hectare. Yield potential is lower when corn is planted late. Number of cobs and grains was determined by counting them respectively.

Grain and biomass weight: Grain weight and dry biomass was determined by using electronic analytical balance.

Statistical analysis

Data collected were subjected to analysis of mean and standard deviation by using MS Excel and SPSS. It is a non-restrictional design in the sense that there is no blocking and any treatment can be replicated any number of times. Five maize plants are selected randomly from the lysimeter and agricultural land as control group and experimental group respectively. Such random assignment of items to two groups is technically described as principle of randomization. Thus, this design yields two groups as representatives of the population. As the experimental units are homogenous, completely randomized design (CRD) has been used for the data analysis and interpretation of results.

Source of Inputs

Seed, fertilizers, and manure are the major inputs used for maize. The Agricultural Inputs Corporation (AIC), a public sector undertaking, was the only institution marketing fertilizers until a few years ago, when its monopoly ceased following changes in government policy. It was envisioned that the private sector would step into supply these inputs, but this has not come to pass. The private sector does supply inputs to Terai districts, but supplies limited quantities in the eastern and central midhills and almost none in the mid-western and far-western midhills and highhills. Some NGOs have been supplying seed, fertilizer, and plant protection chemicals in some areas for vegetable cultivation, but not for maize. Negligible amounts of pesticides are used in maize production in the midhills and highhills. However, pesticide use is common among farmers in the central Terai. All of the pesticide used, especially in the Terai, is purchased from agrovets in nearby markets. Agrovets supply a limited amount of hybrids and improved maize seeds in comparatively accessible areas. Their interest, however, remains on hybrid seed, which has higher profit per unit of seed sold.

Farmyard manure is one of the inputs that every household uses in maize fields. Though farmyard manure is also used for other crops, the largest part of the manure is used for maize production (Paudyal et al., 2001). Soil application of 60:30:0 kg NPK/ha is used to get high yield (Dhital et al., 1990). The grain yield increases significantly with the increasing dose of nitrogen (Tripathi and Pathak, 1984). The chemical fertilizer especially nitrogen fertilizer is universally accepted as a key component to higher corn yield and optimum economic return (Gehl, et al., 2005). The loss of soil fertility and lower use of fertilizer input is also another important factor responsible for low yield of maize. The amount of nitrogen to be applied depends largely on the plant population/unit area of land. Optimum plant population can result in increased production only if there is proper supply of nutrients, particularly nitrogen. Srivastav et al., (2002) found non-significant response of early and late varieties to three levels of N (60, 90 and 120 kg ha⁻¹) and plant densities (53000, 71000 and 95000 plants ha⁻¹) during summer but a significant response during winter season. Adhikari, et al., (2004) reported that the highest grain yield of 9352 kg ha⁻¹ was produced when the crop was fertilized with 120 kg N ha⁻¹ on the crop planted under the plant population of 53,333 plants ha⁻¹; and they noted the lowest yield (6657 kg ha-1) with the crop supplied with 60 kg N ha⁻¹ under plant population of 44,444 plants ha⁻¹.

Problems encountered

1- Southern leaf blight, a new disease of maize is a great threat in the Terai. This disease has reduced plant height, grain yield, biomass and grain weight.

2- Army worm, blister beetle, cut worm, moth, stem borer, termite, weevils, white grub etc are the agents causing damage to maize at different stages in Terai.

3- Erratic rainfall

4- Drought is prolonging year after year in Nepal. Development of drought tolerant varieties is the answer to the drought problem. Adjustment of planting season could help to escape the critical stage of crops to escape from the drought.

RESULTS AND DISCUSSIONS

The results obtained from this study shows that when the maize is given its full water requirement, 486.6 mm of water is required. The maximum amount of water i.e. 318 mm is required and utilized at the mid stage (flowering and fruiting). However, for higher yields, crop water requirement is 500-600mm depending upon the climate and duration of the crop, there should be adequate water during the crop establishment period (Nayava and Gurung, 2010).

In this study, K_c is 0.11, 0.35, 1.51 and 0.34 for initial, development, mid-season and the late season stages, respectively as indicated in Table 1. However, values of Kc depend on the month and locality. The range of monthly values is 0.50-0.80. Average value of Kc for the season for maize is 0.65 (Subramanya, 2007). Similarly, the crop coefficient value was found to be higher as the number of days increases. The

Table 1: PET, T, R, Kc and AI

Months/Stages	R (mm)	PET (mm)	Kc	T in °F	AI
May/Initial	44.00	24.00	0.11	85.35	1.83
June/Crop development	202.10	74.10	0.35	87.04	2.73
July/Mid	429.90	317.90	1.51	85.47	1.35
August/Late	106.60	70.60	0.34	86.43	1.51

Potential evapotranspiration, PET; Temperature, T; Rainfall, R; Crop coefficient, Kc and Aridity index, AI

PET increases as increase in the amount of rainfall and temperature and vice versa as shown in figure 2 and figure 3. However, the PET at mid season has been observed maximum due to the increase amount of rainfall and decrease in temperature. The AI increases with the increase of temperature and observed to be maximum at the crop development stage and minimum at mid stage as shown in table 1.

The crop coefficient (K_c) is affected by a number of factors, which include: the type of crop, stage of growth of the crop and the cropping pattern (Allen *et al.*, 1998). Doorenbos and Pruitt (2000) indicated that plant height and total growing season influence crop coefficient values. The higher the plant height and the longer the growing season the higher the crop coefficient values and vice versa.

Yield components

The relationships between crop yield and water use are complicated. Yield may depend on the timing of water application or on the amount applied. Information on optimal scheduling of limited amounts of water to maximize yields of high quality crops are essential if irrigation water is to be used most efficiently (Anac *et al.*, 1997). Timing, duration and the degree of water stress all affect crop yield.

In this study, height, number of grains, 1000 grains weight and biomass of maize grown in a lysimeter was found to be greater than that of field maize as shown in Table 2. The average yield of maize is estimated to be 1044 kg ha^{-1} when amount of water is provided enough. About 50% reduction in the yield (580 kg ha^{-1}) is observed when there is the scarcity of water. However, the number of grains of maize in a lysimeter and field was negatively affected by the emergence of southern leaf blight disease and other insects and could not reach maximum.

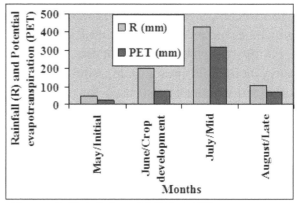

Figure 2: Rainfall and PET at different stages of maize growing period

Table 2: Different parameters of Maize

Parameters	Field	Lysimeter
Average plant height (cm)	218	236.2
No. of cobs	4	3
Width of plant with cobs (cm)	6.5	5.75
Grains of a plant/cob	162	216
1000 grains weight (gm)	160	281
Biomass (gm)	649.9	716.99

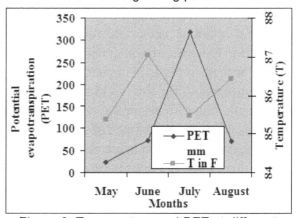

Figure 3: Temperature and PET at different stages of maize growing period

CONCLUSION

Crop coefficient is 0.11, 0.35, 1.51 and 0.34 for initial, development, mid-season and the late season stages respectively. So the range of monthly values of crop coefficient of maize is found to be 0.11-1.51. Potential evapotranspiration is 24 mm, 74.1 mm, 317.9 mm and 70.6 mm for initial, development, mid-season and the late season stages respectively. The total PET is 486.6 mm. The PET of maize is more dependent on rainfall and temperature besides other factors. The PET increases as the amount of rainfall increases. Maintenance of sufficient moisture in a lysimeter has resulted higher yield and therefore the yield of maize can be increased by providing sufficient moisture and manure, controlling disease and pests during initial, crop development and mid season.

ACKNOWLEDGEMENT

The author would like to thank to the director of NWRP, Bhairahawa who gave permission to carry out this research. The author wishes to thank Mr. Ramesh Bhandari, Mr. Dayanidhi Pokharel and his assistants for their cooperation during the whole research period. The facilities and technical assistance provided by the EPTRI, Kathmandu are highly acknowledged.

REFERENCES

1- Adhikari, B. H., Sherchan, D. P. & Neupane, D. D. (2004). Effects of nitrogen levels on the production of maize (*Zea mays* L.) planted at varying densities in Chitwan valley. In: D. P. Sherchan, K. Adhikari, B. K. Bista and D. Sharma (eds.) Proc. of the 24th National Summer Crops Research Workshop in Maize Research and Production in Nepal.

2- Allen R, Pereira L, Raes D, Smith M. (1998). Crop evapotranspiration: guidelines for computing crop water requirements. FAO Irrigation and Drainage Paper No. 56, FAO, Rome, Italy.

3- Anac M.S., Ali Ul M., Tuzal I.H., Anac D., Okur B. and Hakerlerler H. (1999). Optimum irrigation schedules for cotton under deficit irrigation conditions. In: C. Kirda, P. Moutonnet, C. Hera, D.R Nielsen, (Eds.). Crop Yield Response to Deficit Irrigation. Dordrecht, the Netherlands, Kluwer Academic Publishers. Pp. 196-212.

4- Coughlan, M.J. (2003). Defining drought: a meteorological viewpoint. Science for Drought. In: Proc. of the National Drought Forum. Brisbane, Australia, Pp. 24-27.

5- Dhital B.K., Sthapit B.R., Joshi K.D., Pradhanang P.M., Subedi K.D., Vaidya A., Gurung G and Kadayat K.B. (1990). A Report on Maize Research at Lumle Agricultural Center (LAC) and its Off-Station Research (OSR) Site (1989/90) Seminar paper No. 13, Lumle Agricultural Centre, Pokhara, Kaski.

6- Doorenbos J. and Pruitt W.O. (2000). The mechanism of regulation of 'Bartlett' pear fruit and vegetative growth by irrigation withholding and regulated deficit irrigation. Journal of American Society of Horticultural Science. 111: 904.

7- Doorenbos, J. and Pruitt, W.O. (1977). "Guidelines for predicating crop water requirements", FAO Irrigation and drainage paper No.24, FAO, Rome.

8- Gautam, A.K. and Pokhrel, S. (2010). Climate change effects on agricultural crops in Nepal and adaption measures. Presented in Thematic Working Group (agriculture and food security) meeting, Feb 23rd, 2010, Kathmandu, Nepal.

9- Gehl, R.J., Schnidt, J.P., Maddux, L.D. & Gordon, W. B. (2005). Corn yield response to nitrogen rate and timing in sandy irrigated soil. Agron. J., 97:1230-1238. June 28-30, 2004 at NARC, Khumaltar, Nepal, Pp: 216-219.

10- Gurung, G. (2008). Impacts of Climate Change: Some field observations.

11- Hare, F.K. (1993). Climate variation, drought and desertification. WMO, Geneva, Switzerland, Pp. 44.

12- MoAC (Ministry of Agriculture and Co-operatives). (2007). Statistical Information on Nepalese Agriculture (Time Series Information). MoAC, Singha Durbar, Kathmandu, Nepal, Pp. 380.

13- MoE (Ministry of Environment). (2010 a). National Adaption program of action (NAPA) to climate change (report). Ministry of Environment, Kathmandu, Nepal.

14- MoE (Ministry of Environment). (2011). Status of Climate change in Nepal. Ministry of Environment. Government of Nepal.

15- Nayava, J.L., and Gurung, D.B. (2010). "Impact of Climate Change on Production and productivity: A Case study of Maize research and development in Nepal", The journal of Agriculture and Environment Vol: 11, Ministry of Agriculture and Cooperatives, Government of Nepal.

16- Paudyal, D.C., Manandhar, G., Koirala, K.B. (2001). Maize in Nepal: Production Systems, Constraints, and Priorities for Research. Kathmandu: NARC and CIMMYT, Pp. 56.

17- Perry, A. H. (1986). Precipitation and Climate Change in Central Sudan. In: H.R.J. Davies (ed) Rural Development in the White Nile Province, Sudan. A Study of Interaction between Man and Natural Resources, The United Nations University, Tokyo, Pp. 33-42.

18- Practical Action (2009). Temporal and spatial variability of climate change over Nepal (1976-2005). Pratical Action, Kathmandu Office.

19- Rajbhandari, G.R. (1982). A review on various technological aspects of maize. Proceedings of the Ninth Summer Crops Workshop. Parwanipur, Bara, Nepal, Pp. 60-68

20- Shrestha, J., and Timsina, K.P. (2011). "Agronomic Evaluation and Economic Analysis of Winter Maize under Different Plant Population and Nitrogen Rates in Chitwan, Nepal" Nepalese Journal of Agricultural Sciences. Vol. 9, Pp: 5-13.

21- Subramanya, K. (2007). "Engineering Hydrology" Tata McGraw –Hill Publishing Company Limited, New Delhi, India, Pp 369.

22- Tripathi, B.P., and Pathak, L.R. (1984). Response of maize varities to different levels of nitrogen under upland rainfed conditions. Paper presented at the 12th Summer Crops Workshop, Rampur, Chitwan, Nepal.

23- WMO (1994). Guide to Hydrological Practices: Data acquisition and processing, analysis, forecasting and other applications. WMO-No. 168. World Meteorological Organization, Geneva, Pp. 735.

Permissions

All chapters in this book were first published in IJAMAD, by Islamic Azad University; hereby published with permission under the Creative Commons Attribution License or equivalent. Every chapter published in this book has been scrutinized by our experts. Their significance has been extensively debated. The topics covered herein carry significant findings which will fuel the growth of the discipline. They may even be implemented as practical applications or may be referred to as a beginning point for another development.

The contributors of this book come from diverse backgrounds, making this book a truly international effort. This book will bring forth new frontiers with its revolutionizing research information and detailed analysis of the nascent developments around the world.

We would like to thank all the contributing authors for lending their expertise to make the book truly unique. They have played a crucial role in the development of this book. Without their invaluable contributions this book wouldn't have been possible. They have made vital efforts to compile up to date information on the varied aspects of this subject to make this book a valuable addition to the collection of many professionals and students.

This book was conceptualized with the vision of imparting up-to-date information and advanced data in this field. To ensure the same, a matchless editorial board was set up. Every individual on the board went through rigorous rounds of assessment to prove their worth. After which they invested a large part of their time researching and compiling the most relevant data for our readers.

The editorial board has been involved in producing this book since its inception. They have spent rigorous hours researching and exploring the diverse topics which have resulted in the successful publishing of this book. They have passed on their knowledge of decades through this book. To expedite this challenging task, the publisher supported the team at every step. A small team of assistant editors was also appointed to further simplify the editing procedure and attain best results for the readers.

Apart from the editorial board, the designing team has also invested a significant amount of their time in understanding the subject and creating the most relevant covers. They scrutinized every image to scout for the most suitable representation of the subject and create an appropriate cover for the book.

The publishing team has been an ardent support to the editorial, designing and production team. Their endless efforts to recruit the best for this project, has resulted in the accomplishment of this book. They are a veteran in the field of academics and their pool of knowledge is as vast as their experience in printing. Their expertise and guidance has proved useful at every step. Their uncompromising quality standards have made this book an exceptional effort. Their encouragement from time to time has been an inspiration for everyone.

The publisher and the editorial board hope that this book will prove to be a valuable piece of knowledge for researchers, students, practitioners and scholars across the globe.

List of Contributors

Ali Bagherzadeh
Assistant professor of Economics, Khoy Branch, Islamic Azad University, Iran

Fatemeh kazemzadeh
MA in Economics, Tabriz Branch, Islamic Azad University, Iran

Yudi Ferrianta
Department of Agriculture Economic, University of Lambung Mangkurat, Indonesia

Nuhfil Hanani, Budi Setiawan and Wahib Muhaimin
Department of Agricultural Economic, University of Brawijaya, Indonesia

Stephen Umamuefula Osuji Onyeagocha
Department of Agricultural Economics, Federal University of Technology, Owerri, Imo State

Sunday Angus Nnachebe Dixie, Chidebelu and Eugene Chukwuemeka Okorji

Agom, Damian Ila, Susan Ben Ohen and Kingsley Okoi Itam
Deptartment of Agric Economics, University of Calabar, Calabar, Nigeria

Nyambi N. Inyang
Commercial Agric Development Project (CADP), Cross River State, Nigeria

Mafimisebi Taiwo Ejiola
Department of Agricultural Economics and Extension, The Federal University of Technology, Akure, Nigeria

Okunmadewa Foluso Yinka
World Bank Country office, Asokoro, Abuja, Nigeria

Abura Odilla Gilbert and Barchok Kipngeno Hillary
Chuka University College, P. O. Box 109, Chuka, Kenya

Onyango Christopher Asher
Egerton University, P. O. Box 536, Egerton, Kenya

Morteza Tahami Pour and Mohammad Kavoosi Kalashami
Ph.D Student of Agriculture Economics, Department of Agriculture Economic, University of Tehran, Iran

Iniobong A. Akpabio, Nsikak-Abasi A. Etim and Sunday Okon
Department of Agricultural Economics and Extension, University of Uyo, PMB 1017, Uyo, Akwa-Ibom State, Nigeria

Seyyed Ali Noorhosseini-Niyaki
Department of Agronomy, Lahidjan Branch, Islamic Azad University, Lahidjan, Iran

Mohammad Sadegh Allahyari
Department of Agricultural Management, Rasht Branch, Islamic Azad University, Rasht, Iran

Ejaz Ashraf and Abu Bakar Muhammad Raza
Assistant Professor, Agriculture. Extension Education. University College of Agriculture, University of Sargodha, Punjab- Pakistan

Samiullah
Lecturer in Statistics, University College of Agriculture, University of Sargodha, Punjab-Pakistan

Muhammad Younis
University College of Agriculture, University of Sargodha, Punjab-Pakistan

Mohammad Kavoosi Kalashami
Lecturer, Islamic Azad University, Rasht Branch, Rasht, Iran

Morteza Heydari and Houman Kazerani
MSc students of Agricultural Management, Islamic Azad University, Rasht Branch, Rasht, Iran

Seyed Safdar Hosseini
Professor, Department of Agricultural Economics, College of Agriculture, University of Tehran,Karaj,Iran

Afsaneh Nikoukar
Assistance professor, Payam-e-noor University, Khorasan Razavi, Mashhad, Iran

Arash Dourandish
Assistance professor, Department of agricultural economics, College of Agriculture, Ferdowsi University of Mashhad, Iran

Reza Movahedi
Assistant Professor Agricultural Extension and Education Department, Bu-Ali- Sina University, Hamedan, Iran

Masoud Samian and Khalil Mirzai
M.Sc. Student, Agricultural Extension and Education Department, Bu-Ali Sina University, Hamedan, Iran

Adel Esmaili Saloomahalleh
M.Sc. Student, English Language Education Department, Islamic Azad University, Takestan Branch, Takestan, Iran

Bhaskar Pant
Assistant Manager, Axis Bank Ltd

Alpa Rathi
Deputy Manager, National Commodity & Derivatives Exchange Ltd

Anshul Rathi
Senior Research Scholar, College of Technology and Engineering

Mohammad Hossein Bijeh keshavarzi
Young Researchers Club, North Tehran Branch, Islamic Azad University, Tehran, Iran

Mary S. Holz-Clause
Vice President for Economic Development, University of Connecticut, 304 Gulley Hall, Storrs, CT 06268, U.S

Vikram Swaroop Chandra Koundinya
Postdoctoral Research Associate, ISU Extension and Outreach, 220 Curtiss Hall, ISU, Ames, IA 50011, U.S

Nancy K. Franz
Program Director for Families, ISU Extension and Outreach, 111 MacKay, Ames, IA 50011, U.S

Timothy O. Borich
Program Director for Communities, ISU Extension and Outreach, 126 Design, Ames, IA 50011, U.S

Govinda Bhandari
Environment Professionals' Training and Research Institute Pvt. Ltd. Kathmandu, Nepal

Printed in the USA
CPSIA information can be obtained
at www.ICGtesting.com
JSHW051447221024
72173JS00006B/1598